Contemporary Issues in Science Communication

Series editor: **Clare Wilkinson**, University of the West of England, Bristol

---

As science communication continues to establish itself as a discipline in the 21st century, this book series links the present with the past to develop our understanding of its history, both in practice and as an academic discipline.

Out now in the series:

Queering Science Communication;
Representations, Theory, and Practice
Edited by **Lindy A. Orthia and Tara Roberson**

Find out more at
bristoluniversitypress.co.uk/
contemporary-issues-in-science-communication

CONTEMPORARY ISSUES IN SCIENCE COMMUNICATION

RACE AND SOCIOCULTURAL INCLUSION IN SCIENCE COMMUNICATION

Innovation, Decolonisation, and Transformation

Edited by
Elizabeth Rasekoala

First published in Great Britain in 2023 by

Bristol University Press
University of Bristol
1–9 Old Park Hill
Bristol
BS2 8BB
UK
t: +44 (0)117 374 6645
e: bup-info@bristol.ac.uk

Details of international sales and distribution partners are available at bristoluniversitypress.co.uk

British Library Cataloguing in Publication Data
A catalogue record for this book is available from the British Library

ISBN 978-1-5292-2679-9 hardcover
ISBN 978-1-5292-2681-2 ePub
ISBN 978-1-5292-2682-9 ePdf

Cover design: Liam Roberts Design
Front cover image: Liam Roberts Design
Bristol University Press uses environmentally responsible print partners.
Printed in Great Britain by CPI Group (UK) Ltd, Croydon, CR0 4YY

Contents

Series Editor Preface viii
Notes on Contributors x
Acknowledgements xx

Introduction: Race and Sociocultural Inclusion in Science Communication – Global Contemporary Issues 1
Elizabeth Rasekoala

PART I The Practice(s) of Science Communication: Challenges and Opportunities for Race, Gender, Language and Epistemic Diversity, Representation, and Inclusion

1 Inclusion Is More Than an Invitation: Shifting Science Communication in a Science Museum 19
C. James Liu, Priya Mohabir, and Dorothy Bennett

2 Communicating Science on, to, and with Racial Minorities during Pandemics 35
John Noel Viana

3 Breaking the Silos: Science Communication for Everyone 48
Amparo Leyman Pino

4 Building Capacity for Science Communication in South Africa: Afrocentric Perspectives from Mathematical Scientists 63
Mpfareleni Rejoyce Gavhi-Molefe and Rudzani Nemutudi

PART II Science Communication in the Global South: Leveraging Indigenous Knowledge, Cultural Emancipation, and Epistemic Renaissance for Innovative Transformation

5 Challenges of Epistemic Justice and Diversity in Science Communication in Mexico: Imperatives for Radical Re-positioning towards Transformative Contexts of Social Problem-Solving, Cultural Inclusion, and Trans-disciplinarity 85
Susana Herrera-Lima and Alba Sofía Gutiérrez-Ramírez

6	Past, Present, and Future: Perspectives on the Development of an Indigenous Science Communication Agenda in Nigeria *Temilade Sesan and Ayodele Ibiyemi*	100
7	Harnessing Indigenous Knowledge Systems for Socially Inclusive Science Communication: Working towards a 'Science for Us, with Us' Approach to Science Communication in the Global South *Konosoang Sobane, Wilfred Lunga, and Lebogang Setlhabane*	115
8	Indigenous Science Discourse in the Mainstream: The Case of 'Mātauranga and Science' in *New Zealand Science Review* *Ocean Ripeka Mercier and Anne-Marie Jackson*	130
PART III	**The Decolonisation Agenda in Science Communication: Deconstructing Eurocentric Hegemony, Ideology, and Pseudo-historical Memory**	
9	Decolonising Initiatives in Action: From Theory to Practice at the Museum of Us *Brandie Macdonald and Micah Parzen*	149
10	Falling from Normalcy? Decolonisation of Museums, Science Centres, and Science Communication *Mohamed Belhorma*	160
11	African Challenges and Opportunities for Decolonised Research-Led Innovation and Communication for Societal Transformation *Akanimo Odon*	176
12	Decolonising Science Communication in the Caribbean: Challenges and Transformations in Community-Based Engagement with Research on the ABCSSS Islands *Tibisay Sankatsing Nava, Roxanne-Liana Francisca, Krista T. Oplaat, and Tadzio Bervoets*	188
PART IV	**The Globally Diverse History of Science Communication: Deconstructing Notions of Science Communication as a Modern Western Enterprise**	
13	Shen Kua's *Meng Hsi Pi T'an* (*c* 1095 CE): China's First Notebook Encyclopaedia as a Science Communication Text *Ruoyu Duan, Biaowen Huang, and Lindy A. Orthia*	209
14	Making Knowledge Visible: Artisans, Craftsmen, Printmakers, and the Knowledge Sharing Practices of 19th-Century Bengal *Siddharth Kankaria, Anwesha Chakraborty, and Argha Manna*	222

Conclusion: Advancing Globally Inclusive Science Communication – Bridging the North–South Divide through Decolonisation, Equity, and Mutual Learning 239
Elizabeth Rasekoala

Index 249

Series Editor Preface

Clare Wilkinson

As science communication continues to establish itself as a discipline in the 21st century, there has never been a better time to consider contemporary science communication and its practices. Misinformation and digital marketing are changing the context for science journalism. Emerging political eras are altering the way we think about expertise and trust in policy making, as well as the power of protest. Predatory publishing, open access, and the use of social media are presenting novel contexts for researchers to consider in the communication of their work. Meanwhile, there are contemporary scientific and technological developments that are consistently generating important ethical and social questions. Against this backdrop in 2020, we faced one of the greatest global health issues in a generation, with the onset of the COVID-19 pandemic. Concurrently, international events and social movements have drawn heightened attention to questions of inclusion, equity, and the abuse of power.

Contemporary Issues in Science Communication is a multidisciplinary series, welcoming submissions from a wide range of disciplinary areas, and international in outlook. Books contained in the series seek to be engaging, straightforward, and conversational in style, and of interest to science communication practitioners as well as academic audiences.

Books published in the *Contemporary Issues in Science Communication* series consider such science matters and their relationships to communication, engagement, and broader social conversations. The series links the present with the past by publishing titles that develop our understanding of the history of science communication, both in practice and as an academic discipline. However, they also cover a range of topics relevant to contemporary science communication, including, but not limited to: definitions, history, and ethics of science communication; expertise, replication, and trust; interdisciplinary knowledge; ideologies; knowledge and new forms of media; public policy; gaming, Sci-art, and visual communication and inclusivity in science communication. The scope of the series is broad but so are the challenges facing science communicators and public engagement practitioners.

In this book, Dr Elizabeth Rasekoala and the 30 international authors who have provided chapters for her edited collection tackle such testing scenarios head on. In *Race and Sociocultural Inclusion in Science Communication: Innovation, Decolonisation, and Transformation*, we are skilfully guided by both the editor and chapter authors to confront the challenges of sociocultural inclusion where race is concerned in ways that provoke sustainable, meaningful, and impactful change for the science communication sector. In places, this is no easy read, yet nor should it be, when it can be argued that science communication has neglected to appreciate the histories and contributions of so many, particularly those rooted in and drawing on experiences from the Global South.

In this context, the contributors to this edited collection have graciously provided their time, expertise, and perspectives to share both research- and practice-based insights that provoke dialogue and spaces for listening, focused on both implicit and explicit opportunities for sustainable solutions. Drawn together, these chapters create a powerful provocation to consider the ways in which a lack of inclusivity in science communication, where race, ethnicity, and geography is concerned, may not only have practical implications for the field but also seep through its very theoretical foundations and models.

Beyond these insights, the collection is ingrained in the very ethos of this book series, including its aim to highlight both experienced and new authors. As editor, Elizabeth has carefully curated and supported her authors to embrace their visibility, tell their stories with confidence, and navigate a constructive culture of feedback. I thank Elizabeth, not only for the time it has taken her to cultivate this environment in support of the book but also in having confidence that this book series was the right place to publish it. If you are interested in becoming involved in the series personally, please do get in touch. For now, I hope you find this edited collection prompts both insights and reflection, and I am optimistic that it will start to generate some of the sustainable change for the science communication sector, where race and sociocultural inclusion is concerned, for which so many of the contributors passionately argue.

Notes on Contributors

The chapter authors of this book are listed in order of appearance.

C. James Liu is Senior Research Associate at the New York Hall of Science (NYSCI). His studies focus on learning motivation in informal education, development of STEM (science, technology, engineering, mathematics) programmes, and professional development of museum educators.

Priya Mohabir is Senior Vice President of Youth Development and Museum Culture at NYSCI, where she leads the Alan J. Friedman Center for the Development of Young Scientists to diversify the ways that young people see themselves and have access to STEM careers and works with staff across the institution to embed diversity, equity, and inclusion in NYSCI's strategic priorities.

Dorothy Bennett is Director of Creative Pedagogy at NYSCI. She has over 30 years' experience researching gender equity and design-based STEM education and leads the research and development of digital tools, educator and youth professional development programmes, and exhibit experiences at NYSCI to foster curiosity and creative problem-solving for diverse youth and families.

John Noel Viana is a postdoctoral fellow at the Australian National Centre for the Public Awareness of Science at the Australian National University and a visiting scientist at the Responsible Innovation Future Science Platform of the Commonwealth Scientific and Industrial Research Organisation. His research focuses on equity and diversity considerations in precision health research, innovation, and policy. He has written on the importance of engagement with racial minorities in COVID-19 research. He has also published extensively on ethical issues in dementia research and care, media portrayals of novel biotechnologies, and cross-cultural bioethics.

Amparo Leyman Pino is a learning expert specialising in leadership, education, diversity, and inclusion. Since 1994, she has applied her pedagogical experience in the museum world, where she developed content and programmes before moving into institutional leadership and administration roles managing staff and overseeing budgets. After several years in the education field, she co-founded a school in Mexico City using the latest educational philosophies. Most recently, she has been advising museums around the world on community engagement with services like cultural adaptation, content and curriculum development, programme and exhibit creation, and staff training in cultural capacity. Amparo is an active alumnus of the prestigious Noyce Leadership Institute programme, where she has honed her leadership style and engaged with leaders from around the world. She is a member of the respected consortium The Museum Group, and a Fulbright Specialist.

Mpfareleni Rejoyce Gavhi-Molefe is a mathematical scientist and senior manager of the African Institute of Mathematical Sciences (AIMS) House of Science – the science communication and public engagement hub of AIMS South Africa. Before joining AIMS, she was a postdoctoral fellow in mathematics at the University of Alberta, Canada. She received her PhD in mathematics and MPhil in science and technology studies from Stellenbosch University. Her role as manager of the House of Science includes the provision of leadership, direction, and drive on the public engagement landscape; the development of sustainable public engagement frameworks; the provision of capacity building and training for public engagement across diverse platforms and multidisciplinary approaches; and the monitoring, evaluation, and dissemination of learning of public engagement activities and programmes undertaken by AIMS researchers, students, and alumni. From 2018 to 2019, she served as a member of the South African National Advisory Council on Innovation (NACI) and chaired one of the NACI Working Committees – the Transformation in the National Systems of Innovation Programme (NSI). The transformation programme aims to address the challenges of our current and historical legacy related to gender, race, age, disability, and other intersections of inequality through developing innovative monitoring, evaluation, and assessment trends and enhancing the impact of public and private initiatives on transformation in the NSI.

Rudzani Nemutudi is a deputy director of iThemba LABS (Laboratories for Accelerator-Based Sciences) in South Africa. He is in charge of international relations, training, and institutional performance reporting. Before joining iThemba LABS, he was a postdoctoral fellow at the Cavendish Laboratory Semiconductor Physics Research Group in the United Kingdom. In 1997, he received the prestigious Cambridge Commonwealth Trust Scholarship to pursue his research in semiconductor physics at the University of Cambridge, where he completed his PhD in 2001. Dr Nemutudi's experimental research interest in solid state physics spans the specialist fields of condensed matter physics, proximal probe microscopy, accelerator-based ion beam analysis, and low-temperature electron transport measurement on mesoscopic quantum electronic devices grown by molecular beam epitaxial techniques. He has served in various capacities on several scientific organisation committees. He currently serves as the Associate Secretary-General in the Executive Council of the International Union of Pure and Applied Physics (IUPAP) and is the Deputy Representative of IUPAP in the Standing Committee for Gender Equality in Science, which is a global body of international scientific unions striving to promote gender equality and global inclusivity across all fields of science.

Susana Herrera-Lima is a researcher and professor in the Sociocultural Studies Department in ITESO University, in Jalisco, Mexico. She has a PhD in the social sciences with a focus on communication, culture, and society, and a master's in communication of science and culture. She is a member of the National Researchers System. Her research addresses the sociocultural perspective of science communication. Her lines of research are situated in the intersection between public communication of science and environmental communication. Susana is an advisor for the development of social problems-driven projects in science and environment communication with citizen participation, working with civil society organisations, scientists, and communicators. She heads the Permanent Seminar in Water Studies in ITESO. Susana is founder and coordinator of the book series *De la academia al espacio público: Comunicar ciencia en México (ITESO)*. She is a member of international and national academic networks, such as the International Environmental Communication Association (IECA), the International Network on Public Communication of Science and Technology (PCST), Latin America and the Caribbean Science Communication Network (RedPOP), Sociedad Mexicana para la Divulgación de la Ciencia y la Técnica [Mexican Society for the Popularization of Science and Technology] (SOMEDICYT), and Waterlat. She is a member of the board of the *Journal of Science Communication America Latina* (JCOM AL). She has published books, chapters, and papers in Mexican and international journals.

Alba Sofía Gutiérrez-Ramírez is a chemist turned science communicator with a master's degree in science communication. She is currently in charge of the Science Communication Office at the Institute of Neurobiology, National Autonomous University of Mexico. Her science communication work combines outreach, research, and communication training. She collaborates on several research projects regarding science communication, social action, and social justice, particularly with vulnerable communities in Mexico. She has also been a consultant and project manager for knowledge and technology transfer projects, technological innovation, and university–industry collaborations. She is a member of the PCST Network and the Mexican Society for Science and Technology Popularisation. She strives to make science communication more culturally relevant, accessible, and pertinent for social action.

Temilade Sesan is a sociologist with ten years' research and consulting experience in international development. Her research examines the socio-economic dimensions of the food–energy–waste nexus in West Africa (particularly Nigeria and Ghana), especially as these dimensions intersect with issues of the environment, gender, health, and urban planning. She works across these sectors to identify pathways to greater inclusion of marginalised groups – including women and girls – in public and private development initiatives. A key focus of her work is highlighting the social and cultural upheavals that often accompany seemingly benign processes of technological and economic change in the region. Dr Sesan works with a range of policy and development actors to generate evidence-informed and actionable knowledge outputs, thereby amplifying the societal impact of her research. She has communicated her research through peer-reviewed scientific articles as well as blogs, documentaries, opinion pieces, and policy briefs. She teaches and supervises postgraduate students at the University of Ibadan, Nigeria.

Ayodele Ibiyemi is a researcher, critic, and writer from Nigeria. He has consulted for media, research, and not-for-profit organisations. He was awarded his bachelor's and master's degrees from Obafemi Awolowo University, Ile-Ife, Nigeria. He works at the intersection of open governance, language advocacy, and Indigenous knowledge development and has contributed to a number of volumes on these topics. In 2019, he won the Ken Saro-Wiwa Prize for Critical Review and currently works at Marquette University's Indigeneity Lab.

Konosoang Sobane is Chief Research Specialist in Science in Society at the Human Sciences Research Council (HSRC), South Africa. She is also a research associate of the Department of Strategic Communication at the University of Johannesburg. She holds a PhD in linguistics from Stellenbosch University and a postgraduate diploma in science communication. She has worked intensively in the fields of strategic communications, corporate communications, and science communication. Her research interests are science communication, effective utilisation of digital communication, translation and communication of research, social and behavioural change communication, and public engagement.

Wilfred Lunga is a multidisciplinary researcher, disaster risk reduction expert, and educationist with a development, vulnerability reduction, climate change, and resilience-strengthening background in the social sciences at the HSRC, South Africa. His strength as a researcher is in networking, fundraising, management of large-scale surveys, and the facilitation of diverse groups and successfully managing to stimulate discussions with opposing views within groups.

Lebogang Setlhabane is currently working at HSRC, South Africa, as a research assistant and a junior researcher. Lebogang received her national diploma, baccalaureus, and magister technologiae in language practice from the Tshwane University of Technology (TUT). She was awarded a TUT postgraduate scholarship that funded her magister technologaie studies and is currently registered for her doctor of language practice at TUT specialising in sociolinguistics.

Ocean Ripeka Mercier (Ngāti Porou) works at Te Kawa a Māui (the School of Māori Studies) at Victoria University of Wellington, Aotearoa New Zealand. She has a PhD in materials physics. Her teaching and research examine the connections between mātauranga Māori (Indigenous Māori knowledge) and science, and how these can contribute to healthier natural environments – through pest control, decolonising and indigenising marine management, and better understanding of groundwater systems. She is a presenter on TVNZ's Coast New Zealand and the presenter of Māori Television's science show Project Mātauranga. Her work in science communication saw her receive the New Zealand Association of Scientist's Cranwell Medal in 2017 and the Royal Society of New Zealand: Te Apārangi Callaghan Medal in 2019.

Anne-Marie Jackson (Ngāti Whātua, Ngāti Kahu o Whangaroa, Ngāpuhi, Ngāti Wai) is co-director of Te Koronga Indigenous Science Research Theme and Graduate Research Excellence programme at the University of Otago. She is also a co-director of Coastal People: Southern Skies, one of Aotearoa New Zealand's Centres of Research Excellence. She has a doctorate in Māori studies and physical education, examining rangatiratanga and Māori health and well-being within a customary fisheries context. Her kaupapa (the collective philosophy principle) is mauri ora (flourishing wellness) and she focuses on hauora (Māori physical education and health), Tangaroa and the marine environment, waka and water safety, and Indigenous science.

Brandie Macdonald (she/her; Chickasaw/Choctaw) is Science Communication Consultant and the former Senior Director of Decolonising Initiatives at Museum of Us and a PhD fellow in Education Studies at the

University of California in San Diego. Her work as a practitioner and researcher focuses on the transformative systemic change in museums that is driven by anticolonial and decolonial theory-in-practice, and how the symbiotic application of decolonial policy and practice can redress colonial harm domestically and internationally.

Micah Parzen (he/him) is a husband, a father, an anthropologist, and an attorney. He joined the Museum of Us as its CEO in August 2010. He and his team have since transformed the museum from a musty, dusty, and tired institution into a cutting-edge museum. Micah currently serves on several boards that work towards supporting the San Diego community and beyond.

Mohamed Belhorma is a museum professional in scientific outreach and communication. He currently creates exhibitions and outreach content around physics and future energy issues. He holds an MA in art history curatorial studies and an MA in science museology. Mohamed's work has focused on the practical aspects of designing science communication content. His current project uses interaction design solutions to help reduce the impact of sociocultural barriers.

Akanimo Odon holds a master's degree in environmental rehabilitation, a PhD in environmental management from Lancaster University, UK, and business and enterprise certifications and fellowships from Cambridge University, UK, and Stanford University, United States. He specialises in cross-border education and innovations in research and is an expert in navigating, developing, and managing relationships between academia, government, and industry in Africa for economic viability, research impact, and student employability. He has been Africa Strategy Advisor for Aberystwyth University, University of Strathclyde, and the University of East Anglia and is currently the African Regional Advisor for the University of London, Lancaster University, and the Zurich Elite Business School, in Switzerland. He recently founded 'Flexy-Learn' (www.flexylearn.com), a one-stop online shop for distance learning provision for Africa. He is Principal Consultant of Envirofly Consulting UK (www.envirofly-group.com) and founder of Lancaster International, an international outreach school/programme focused on supporting primary and secondary schools in Africa (www.lancasterinternational.org). He is also currently Director of African Partnerships for Lancaster University and Chair of the African Union Scientific Research and Innovation Council UK Diaspora Engagement Chapter.

Tibisay Sankatsing Nava (she/her) is a PhD candidate and community engagement coordinator from Aruba. She has worked in Dutch academia for ten years and currently works at the Royal Netherlands Institute of Southeast Asian and Caribbean Studies in the Netherlands as part of research programmes focused on Caribbean islands, such as Island(er)s at the Helm and CaribTRAILS. She also works with the Dutch national funding agency to strengthen research infrastructures in the ABCSSS islands. These experiences have been fundamental to her embodied awareness of the inequities in research and of the role of academic research in the Caribbean. This has led her to an ongoing critical reflection on her own ambiguous position within Dutch academia. Tibisay's research and practice focus on this ambiguity and on the co-creation of research with Caribbean island communities.

Roxanne-Liana Francisca works as a biologist at Stichting Nationale Parken (STINAPA), Bonaire. Born and raised in Curaçao, she grew up in and around the sea and has always been interested in conservation efforts and the relationships between people and protected areas and species. She has first-hand experiences with the disconnect between how academia communicates and how Bonairians listen. As one of the few biologists on the island with roots in the former Netherlands Antilles, she is often relied upon to communicate and explain STINAPA's conservation interventions. Roxanne-Liana is interested in increasing the sense of ownership and participation of local people in nature conservation. She continues to work to ensure that all Bonaireans have access to nature information in a way that is digestible to them.

Krista T. Oplaat (she/her) works as an executive secretary and policy officer at Mental Health Caribbean (MHC) on Bonaire. Born and raised in Aruba, to a Colombian mother and Dutch father, the confluence of different cultures has always been her bread and butter. Her fascination with the mysteries of the brain led her to study neurobiology and science communication in Amsterdam. After this, the pull back to the Caribbean and personal experience in the detrimental effects of poor mental health care led her away from a career in (Dutch) academia to tackle the stigma and betterment of mental health on the islands. This drive and her adaptive and socioculturally flexible nature has been integral to her current work in communication and advising on strategy policy at MHC.

Tadzio Bervoets is the former Director of the Dutch Caribbean Nature Alliance, a network organisation that supports nature conservation on all six islands of the Dutch Caribbean. He was born in Sint Maarten to a local Sint Maarten mother and a Belgian father. Tadzio has an undergraduate degree in international relations and NGO management and a graduate degree in environmental resource management. Tadzio worked in Bermuda advising the government on coral reef conservation and in Tanzania doing research and developing incentives for local fishermen to become marine park rangers. He returned to the Caribbean to serve as the marine park manager in Sint Eustatius. Soon after, he became Director for the Sint Maarten Nature Foundation, where he was instrumental in establishing Sint Maarten's first National Park.

Ruoyu Duan currently works at a provincial museum in China. She obtained her master of science communication degree at the Centre for the Public Awareness of Science, Australian National University. Her current research interests are in the history of science communication and informal learning at museum and science centres within historical, cross-cultural, and psychological educational contexts.

Biaowen Huang is Associate Professor in the Department of Communication at Beijing Jiaotong University, China. His research interests include science communication, environmental communication, and health communication, specifically to explore public attitudes and behaviours towards scientific and health issues and to understand information diffusion mechanisms in social media.

Lindy A. Orthia is Honorary Senior Lecturer at the Australian National University School of Sociology. She worked for more than a decade as a science communication teacher and researcher, during which time she published over 40 peer-reviewed works and was awarded five teaching awards. Her research focuses on questions of equity, diversity, and inclusion, and her research interests include science communication histories and representations of science in popular fiction.

Siddharth Kankaria is a science communication practitioner and researcher interested in developing intersectional science engagement practices for the Global South. He currently works as the Communications and Programme Coordinator for the Simons Centre at the National Centre for Biological Sciences in Bengaluru, where he also teaches two graduate-level courses in science communication. Siddharth has a BS (research) in biology from the Indian Institute of Science, Bangalore, and an MSc in science communication and public engagement from the University of Edinburgh, UK, and is also the founder of SciCommSci Club – a flagship initiative for engaging with the science of science communication. He spends his time exploring the research–practice–teaching continuum within science communication; contributing to mentorship, capacity building, Diversity, Equity and Inclusion efforts; and building policy, infrastructure, and communities of practice around science engagement in India.

Anwesha Chakraborty is Research Fellow at the Department of Political and Social Sciences, University of Bologna, Italy. She completed her PhD on Indian science and technology policies and scientific institutions post-independence from the same university in 2017. Anwesha's present research interests lie at the intersection of open data-related practices, grassroots innovation by civil society actors, digital governance, and Science and Technology Studies (STS). Her work has appeared in international journals such as *Journal of Science Communication*, *Asian Studies Review*, *First Monday*, *Telecommunications Policy*, and *Sociologia del Lavoro* (the Italian journal of sociology of labour).

Argha Manna is a science journalist and comics artist from Kolkata, India, and is currently an artist-in-residence at the Humanities and Social Sciences Division at IIT-Gandhinagar. He explores the historical perspectives of scientific development and social issues through the lenses of an artist and a researcher in order to bridge the gap between academic research and visual storytelling. In his latest project, 'Famine Tales from India and Britain', in collaboration with the University of Exeter and the British Library, Argha has weaved a meta-fiction narrative to bring archival materials related to the Great Bengal Famine of 1770 to life through comics. Argha also works as a visual consultant for the CoFutures project as part of the University of Oslo and Values in Design Laboratory at UC Irvine, where he is writing a comics book on ant network theory, co-evolution, and planetarity. During the COVID-19 pandemic, he collaborated with Bourouiba Lab at MIT in translating the complex science of the fluid dynamics of disease transmission into comics-based teaching tools.

Elizabeth Rasekoala is the President of African Gong: The Pan-African Network for the Popularization of Science & Technology and Science Communication, which aims to advance public learning and understanding of science, scientific outreach, and scientific literacy on the African continent.

Elizabeth's journey in science communication began in the UK in the early 1990s, where through her lived experiences in postgraduate study and later professional work, she became a champion and advocate for race and gender equality in STEM education and science communication. Through the African-Caribbean Network for Science and Technology, which she co-founded with other Black STEM professionals, she undertook the delivery of ground-breaking initiatives to advance race and gender equality for Black youth and adults and underserved ethnic minority communities. These initiatives included: The National RESPECT Campaign; Delivering Inclusion in Science Communication; the network of Ishango Science Clubs in UK cities; and the EU-funded project ETHNIC, involving seven EU countries.

With a professional background in chemical engineering, Elizabeth has researched, presented, and written widely on public innovation and transformative development through advancing diversity, sociocultural inclusion, and race and gender equality issues in science communication and STEM education and skills development. She has provided extensive advisory and consultancy expertise to various governments and agencies both in the Global North and South, and to multilateral international organisations over the past 20 years, including the European Commission, the UN Commission on Human Rights, the UN Economic Commission for Africa, UNESCO, the African Union Commission, and the African Development Bank.

Elizabeth has been internationally recognised for her innovative, dynamic advocacy, and transformative practice with various awards, the most recent being the International NAT AWARD for Science Communication – conferred by the Natural Science Museum of Barcelona, in 2019.

Acknowledgements

A book project of this magnitude in its conceptual framework, thematic depth, and geographical scope could not have been produced without the solidarity and support of many colleagues in the field, who are enrolled in its ethos and engaged in advocacy for the many inequalities it highlights and can thus bring their longstanding knowledge, expertise, and good practice on board. They have provided challenging and empowering perspectives, strengthened the insights of chapter contributors, and enhanced the scholarship of this book through their contributions as individual chapter peer-reviewers.

Their contributions to this book project have been invaluable, and they are acknowledged in alphabetical order:

- Ama Biney – University of Liverpool, UK;
- Amelia Bonea – University of Heidelberg, Germany;
- Sally Deffor van der Heijden – DAI Global, Netherlands;
- Noah Weeth Feinstein – University of Wisconsin-Madison, United States;
- Jean S. Fleming ONZM CRSNZ – University of Otago, Aotearoa New Zealand;
- Azeza Fredericks – independent science communicator, South Africa;
- Alphonse Keasley Jr – University of Colorado Boulder, United States;
- Francis Kombe – EthiXPERT and African Research Integrity Network, Kenya;
- Relebohile Moletsane – University of KwaZulu-Natal, South Africa;
- Sujatha Raman – Australian National University, Australia;
- Tafadzwa Tivaringe – Spencer Foundation, United States;
- Samuel J. Ujewe – Canadian Institutes of Health Research, Canada.

To the many colleagues often unheard and unrecognised, working tirelessly to advance diversity, equity, and inclusion in science communication across the globe, we see you, we hear you, and we know you – this one's for you. May it further spur your efforts onwards and upwards.

Finally, acknowledgements to all the supportive personnel at Bristol University Press, who have created an environment for the production of this edited collection that has facilitated and encouraged innovative thinking and transformative approaches: the Contemporary Issues in Science Communication book series co-editors Clare Wilkinson and Paul Stevens; Georgina Bolwell and the production team; and Bahar Celik Muller and the marketing team.

Introduction: Race and Sociocultural Inclusion in Science Communication – Global Contemporary Issues

Elizabeth Rasekoala

The historical and contemporary trajectories of societal change and transformation on socio-economic and sociocultural indicators of inequality – on race, gender, social class, and other parameters – demonstrate that progressive systemic change is neither inevitable nor to be taken for granted and presumed as sustainable in an ever forward-moving direction. Given the cautionary lessons of these historical and contemporary trends, advocates of systemic change where inclusion is concerned should continuously interrogate seemingly progressive developments. They should, furthermore, also engage their fields through solidarity, to assert reflective spaces for the considerations of innovative, globally inclusive, and action-oriented approaches that further advance the pace of change. This is the rationale and ethos that has guided and framed the conceptualisation and development of this edited collection for the science communication field.

The growing engagement with diversity, equity, and inclusion issues in the science communication field, as demonstrated by the programmes of various national, regional, and global conferences, workshops, webinars, and symposiums, are signposts of an emerging field-wide recognition of the pressing need for sociocultural inclusion. Alongside these developments, the increase in science communication journal publications and other platforms by scholars, practitioners, and advocates attests to robust scholarly efforts to engage the field with these challenging issues (Lewenstein, 2019).

Given the global political climate on issues of race and sociocultural inclusion that has emerged as a result of the exigencies of the COVID-19 pandemic, there now seems to be an even greater appetite and interest in insightful knowledge to drive, inform, and advance lasting global change and a willingness to listen to others across the divides of race, region, power, and economic wealth with mutual respect and consideration.

The cautionary aspect here is that the mainstream international discourses on diversity, equity, and inclusion in science communication are in danger of generating much handwringing and concern without seeming to drive consistent change and systematic transformations of the field. This book seeks to elucidate the various 'elephants in the room' of these seemingly circular discourses surrounding diversity, equity, and inclusion and thereby address critical gaps in the literature, practices, capacity building, and scholarship within the science communication field regarding race and sociocultural inclusion. It also brings to the fore the many marginalised voices of so-called racialised minorities and those from Global South regions to further interrogate the globalised footprint of the science communication enterprise. While the themes of this book are very timely in addressing the new paradigms of equity and inclusion advocated for across the globe during the COVID-19 pandemic, societal change and transformation take time, and this collection offers many relevant reflections for the post-COVID-19 era.

The chapters in this book are ambitious in seeking to take current discourses, knowledge, understandings, and momentum for the advancement of diversity, equity, and inclusion in the science communication field to the deeper, multifaceted, and paradigm-shifting levels that should stand a better chance of engendering systematic change (Finlay et al, 2021). This book also uniquely focuses on the current state of the science communication field in the Global South, offering contemporary narratives, insights, and empowering contributions from practitioners and researchers engaged in the field in these 'invisible' parts of the world. Bridging the Global North–South divide in science communication is a critical aim of this collection, which seeks to interrogate and thereby engender consensus as to what diversity, equity, and inclusion mean and should look like across this divide. It is inherent in this divide that race and sociocultural inclusion parameters are critical lenses through which to view and bridge this global divide (Rasekoala, 2016).

The contributing authors from various regions of the world have signposted, showcased, and illustrated innovative, insightful, and transformative opportunities and ways forward for advancing inclusion across these frontiers. Their insights should be catalysts in stimulating new thinking, engendering paradigm-shifting understandings, and creating an impetus for transformative action based on their deep understandings of race and sociocultural inclusion paradigms in the field. The innovative good practice approaches highlighted in the book can serve as inspiring signposts as to how and where transformative and systematic action for change could take the field forward in terms of realising the ultimate goal of a radically transformed and globally equitable landscape for science communication.

This book seeks to enrich the discourses and perspectives of the science communication field through advancing a vista-shifting understanding of what inclusion and equity within science communication contexts should look like from the pivotal perspectives of the very groups that are marginalised and made invisible within it.

It is important at this stage to provide definitions of some of the terms used throughout this book. The need to clarify definitions is crucial in a book of such geographical scope in order to reflect the different growth trajectories of the science communication field at national and regional levels. The term 'Eurocentrism' refers to a globalised worldview that centres Europe and predominantly favours the Western world over other parts of the globe. It stems from the era of the colonisation of much of the world by European nations, and the enduring legacies of the hegemonic dominance of political power, economic wealth, and sociocultural influence that it has accrued to the West at the expense of other regions of the world (Potter, 2018).

In this context, it is useful to explore other worldviews that do not receive the same level of globalised dominance and are aired in this book to enhance inclusive recognitions in a globalised context of increasingly contested worldviews. The first is Afrocentricity – conceptualised by Molefi Asante (2007), who defines it as 'a consciousness, quality of thought, mode of analysis and an actionable perspective where Africans seek, from agency, to assert subject place within the context of African history' (Asante, 2007, p 16). Asante's call to action on Afrocentricity has contributed to many contemporary developments such as the drive for the African Renaissance and the emancipation of African culture and epistemology on the continent and in the African diaspora. The second is Orientalism – conceptualised by Edward Said (1978). Said conceptualised the term Orientalism as a philosophical, scholarly, and advocacy-based critique of East–West relations to describe the stereotypical, prejudiced depictions, and 'exotic' 'othering' of peoples of the East/the Orient – from North Africa, the Middle East, and Asia – by the Western world, throughout history and in contemporary times (Said, 1978).

The terms Global North and Global South have largely grown out of the development sector and in reaction to the growing recognition of global inequalities and the value-laden pejorative terms previously used to differentiate the global development divide – terms such as 'Third World' countries, 'developing' countries (used to denote the regions of Latin America, Asia, Africa, and the Pacific-Oceania) and 'First World'/'developed' countries (used to denote the regions of the Western world – Europe and North America). The use of the more 'progressive' terms of Global North and South is intended to better reflect the growing solidarity of countries and regions in the Global South as they strive to assert their rightful place in the world in the face of the growing global North–South divide of unequal political power, resource allocation, and decision-making (Olatunji and Bature, 2019).

The chapters in this book have been developed and written with the analogy of 'peeling the layers of an onion' in mind. This multifaceted understanding of the interconnectedness of these layers – each layer is fused to the one before and after it – has informed the thematic parts and sub-themes of the collection. Thus, intersections of the paradigms that infuse the diversity, equity, and inclusion challenges in the science communication field are explored, elaborated, interrogated, analysed, challenged and discussed in the chapters of the book via each of its four thematic sections. Furthermore, the book brings to the global fore of the field a range of new and diverse voices and contributions across the young, emerging, and established age and career ranges from diverse multidisciplinary backgrounds: science communicators; environmental activists; researchers; public health practitioners; senior leadership of science centres and museums, managers, curators, and exhibition designers; university academics; non-governmental organisations; science journalists; and individuals from diverse STEM (science, technology, engineering, mathematics) subject disciplines such as mathematical scientists, engineers, biologists, and the social sciences.

These diverse backgrounds and voices have produced a wide variety of different writing and narrative styles across the chapters of the book. These vary from thought-provoking personal expositions of lived experiences as science communication professionals to in-depth case studies describing challenging institutional journeys of transformation through innovative race and sociocultural inclusion good practices. They also include chapters that elaborate academic research outcomes and theoretical concepts and showcase and laud empowering Indigenous and community-based knowledge governance and public engagement practices. These divergent writing styles make this a book collection that is accessible and relatable for the diverse readerships and stakeholders in the science communication field and beyond.

The global scenarios encapsulated by the chapter authors are also very wide ranging. The book has 30 contributing authors (18 female and 12 male) from the following countries: Mexico, the United States, the six Dutch-speaking Caribbean islands (Aruba, Bonaire, Curaçao, Saba, Sint Maarten, and Sint Eustatius – the ABCSSS Islands), the UK, the Netherlands, Germany, Italy, Nigeria, South Africa, India, China, Australia, and New Zealand. The book's global scope is also reflected in the diverse nomenclatures of racial definitions that are used across the chapters to reflect national contexts. The aim is not to achieve consistency but to allow for the different contexts in which the authors are located in the understanding that these descriptions continue to evolve within national and international contexts, as the sensitivities and sensibilities of societies grow and greater awareness of racial diversity and multiculturalism matures across the globe. Some of the various racial nomenclatures used in the book include ethnic minorities, racialised minorities, Black, Indigenous, and Peoples of Colour, and so on. In a similar vein, the terms diversity, equity, and inclusion are aligned interchangeably based on national and regional contexts and preferences, such that in some regions, equity, diversity, and inclusion is used instead, while in others, inclusion, diversity, and equity is preferred.

Part I: The Practice(s) of Science Communication: Challenges and Opportunities for Race, Gender, Language and Epistemic Diversity, Representation, and Inclusion

It has long been recognised that the practices of science communication exclude many sections of society despite the good intentions of its advocates and practitioners (Rasekoala, 2016; Dawson, 2019). The chapters in this part of the book seek to shed light on the insidious nature of these exclusionary practices and the institutional frameworks, structures, and mind-sets that facilitate, maintain, and sustain them, notwithstanding good practice intentions to do otherwise. The authors in Part I also define the radically transformative and systematic journeys that need to be undertaken by science communication institutions and practitioners in order to engender true and lasting diversity, equity, and inclusion in the field.

In Chapter 1, C. James Liu, Priya Mohabir, and Dorothy Bennett elaborate on the ten-year development strategy of their science museum, the New York Hall of Science in the United States, as it undertakes reflective institutional change to transform its science communication culture of inclusion through transforming the race, age, and gender profile of its employees, especially those that interact on the front lines, so that these align with those of its visitors, learners, and diverse communities. These developments also include the engendering of institutional mind-set changes in the dynamics of science communication interactions between these diverse front-line 'Explainers' and the museum's visitors and learners. These transformations have been sustained through a richly empowering training programme, inculcating values, insights, and motivations from the lived experiences of these diverse 'Explainers' themselves.

In Chapter 2, John Noel Viana addresses the current and dynamic issues related to communicating science on, to, and with racial minorities during pandemics. This chapter explores scientific and medical communication during the Severe Acute Respiratory Syndrome (SARS) outbreak in Canada and the COVID-19 pandemic in Australia, which are two countries with significant populations of Asian minorities as well as other racially diverse minorities. The author seeks to encourage a critical interrogation of the nature of what equity, diversity, and inclusion should entail in public engagement with these communities and in the public health and science communication of diseases in times of crises. These considerations of sociocultural inclusion, the chapter argues, are paramount given the systemic inequalities in health disparities and racial discrimination in the lived experiences of these minority ethnic populations.

The dynamic nature of multiracial demographic changes in the growing population diversity of many countries in Global North regions, through historical and contemporary migration patterns, creates an imperative for innovative transformations in the methodologies and approaches of the informal science education sector, such as science centres and museums. The recommendation in Chapter 3 from Amparo Leyman Pino is that of delivering systemic change by moving from the limitations of 'integration tactics', which simply create 'silos' of superficial integration, to the multidimensional lenses of 'inclusion strategies'. The rationale here is of good practice strategies that are powered by the leveraging of 'mission-driven intentionality', 'blended language' programming, universal design principles, equitable collaborative partnerships with minority communities, and fully addressing the 'chain of accessibility' for excluded groups within and beyond institutional settings and contexts.

In Chapter 4, Mpfareleni Rejoyce Gavhi-Molefe and Rudzani Nemutudi address the capacity-building challenge of growing the inclusive and diverse human resource for science communication in South Africa despite a well-developed national science and science communication policy framework for transformation. Inherent in this challenge is the pernicious legacy of colonialism and apartheid, which has systematically excluded much of the majority Black African population from science education, research, and communication practices, opportunities, and benefits. It signposts an innovative Afrocentric science communication capacity-building initiative that has been delivered since 2020 in South Africa, with transformational impacts, for young and emerging African researchers in mathematical sciences – the Africa Scientifique: Leadership, Knowledge, and Skills for Science Communication Programme.

Part II: Science Communication in the Global South: Leveraging Indigenous Knowledge, Cultural Emancipation, and Epistemic Renaissance for Innovative Transformation

The chapters in Part II seek to illustrate and provide exemplars and thought leadership on the current state of the science communication field in the Global South regions of the world and the challenges faced in overcoming the legacies of Eurocentric hegemony in their postcolonial societies, particularly in advancing the scientific enterprise. In essence, at the heart of this challenge is the imperative of the wholesale liberation of science communication across the Global South from the Eurocentric dominance of knowledge systems, epistemology, and cultural references within institutional frameworks and practices (see Quijano, 2000; Seleti, 2012; Hikuroa, 2017; Wade, 2019; Finlay et al, 2021). Thus, the chapters here seek to elaborate on how science communicators, researchers, academics, and science journalists in the Global South can unshackle their systems of science capacity building, public engagement practices, and community outreach approaches from these inequitable historical legacies that continue to undermine their good intentions, agency, and equitable transformation agendas.

In addition to these historical legacies, there is the creeping impact of globalisation in its myriad forms across the science communication landscape, leading some to question whether the globalisation of science communication is simply exporting Eurocentrism in the guise of internationalisation. The science communication field is increasingly lauding the growing global footprint of its enterprise, but parallel to this, critical questions need to be addressed with regard to the hegemonic drivers of this growth. It is crucial to interrogate whether this international expansion has been achieved at the profound cost of cultural diversity, localisation of contexts, language plurality, and other critically unique, transformative, and empowering approaches across the regions of the world (Gerber, 2018). The implications of these globalisation trends are profound for the global development of the science communication field, not least for the regions of the Global South, in overcoming the 'Trojan-horse' of Eurocentrism, inherent in these new guises of the internationalisation of the field.

An important case in point is the ubiquitous design, content, exhibits, materials, activities, narratives, and so on of science centres and science museums across the globe, which raise profound questions about the lack of creativity and innovation in the field itself. When science centres and museums in Africa, Asia, Latin America, and across the Global South look and feel like carbon copies of those in Europe and North America, we are then faced with the systematic challenge of the twisted manifestations of globalisation in the field of science communication (Rasekoala and African Gong, 2016).

The chapters in Part II deliver novel pathways for overcoming these hegemonic and inequitable manifestations of globalised science communication in the Global South regions of the world. In Chapter 5, Susana Herrera-Lima and Alba Sofía Gutiérrez-Ramírez recommend the radical re-positioning of science communication in Mexico, shifting the focus of its agenda from the brokerage of scientific knowledge to that of addressing complex social problems collectively with the social actors that are directly affected by them. The authors rationalise that these novel approaches are vital in overcoming the 'coloniality of knowledge' legacies that undermine emancipative epistemic renewal efforts, both at institutional and practice levels, across the Latin American region. In Chapter 6, Temilade Sesan and Ayodele Ibiyemi illuminate the past, present, and future contexts and contestations with regard to the development of a localised Indigenous science communication agenda in Nigeria. They further affirm the drive for the delivery of inclusive language plurality and diversity as inherent for the advancement of transformative science communication in these Global South regions.

Chapter 7 by Konosoang Sobane, Wilfred Lunga, and Lebogang Setlhabane suggests the leveraging of the Indigenous knowledge of science communicators and social actors in the Global South is a significant transformative shift for science communication in these regions to undertake, as well as an impactful means of enhancing local agency, localised contexts, and cultural empowerment. Intriguingly, their chapter postulates that the privileging of Eurocentric science communication in the Global South regions at the expense of their local contexts and Indigenous knowledge should be considered as yet another example of the much problematised 'deficit model' practice in science communication. The authors then use diverse case studies from across the Global South to illustrate how this deficit model can, should, and increasingly is being overturned in these regions by empowered local practitioners.

Chapter 8 by Ocean Ripeka Mercier and Anne-Marie Jackson on Aotearoa New Zealand powerfully centres the Indigenous knowledge of the Māori population as an integral knowledge governance system in and of itself that should be accorded equity in recognition and 'mainstreaming' in science communication in that nation. They further illustrate the transferability of this good practice across other regions, for the inclusion of Indigenous populations worldwide. A key thread that runs across the discourses in these chapters is that of the need for language inclusion in science communication practices across the Global South regions, given the hegemony of colonial-era European languages, such as English, French, Spanish, and Portuguese, at the expense of the many local and Indigenous languages in which are embedded the cultural sensitivities and sensibilities of the peoples of these regions (see Marquez and Porras, 2020; Kago and Cisse, 2022).

Part III: The Decolonisation Agenda in Science Communication: Deconstructing Eurocentric Hegemony, Ideology, and Pseudo-historical Memory

The decolonisation agenda is one that has become much more prominent in science communication discourses over recent years. The chapters in

Part III, however, seek to introduce more multidimensional perspectives beyond the narrow confines of the current discourses that are fixated on the repatriation of objects and artefacts housed in museums in countries of the Global North that were taken from countries in the Global South during the colonial era (Woldeyes, 2019).

These chapters elaborate and frame the pressing issues of the decolonisation agenda in much more nuanced and multifaceted levels and frameworks such as: the problematic nature of science and society connections within the context of un-decolonised museums; considerations of race, cultural diversity, and inclusion and what these mean with regard to the 'objectification' of artefacts from non-Western cultures/backgrounds; the interrogations of the 'entitlement' of the subjective Western gaze as it views these objects from non-Western cultures/backgrounds; and the decolonisation of curation practices.

In addition, these chapters address the decolonisation discourse in science communication as needing to engender the re-visioning and transformative agenda-setting of the science communication field and its practices, narratives, and ethics. These encompass the dynamics of exploring science communication within the realms of the ethical dimensions of its decolonisation practices with regard to addressing issues of 'ownership', belonging, access, equity, and the historical and contemporary perspectives of scientific knowledge (Rasekoala, 2020).

Thus, Chapter 9 by Brandie Macdonald and Micah Parzen makes an important contribution on the role of individuals in senior leadership positions working within colonially constructed museums and how they initiate, up-scale, and sustain the processes of decolonising such institutions. This chapter uses a case study narrative to powerfully illustrate the personal commitments that then drive a transformative conceptual explanation of the meaning of decolonisation as a verb, a reflective process, and an unfolding engagement that is sustained through concerted 'unyielding commitment to continuous action in a forward-moving process of reckoning with the ever-present traumatic legacy of colonialism'.

In Chapter 10, Mohamed Belhorma argues that racial grammar (Said, 1993) deeply structures and embeds Western culture in science communication practices, which then facilitates and sustains the creation of a 'normative' supremacist narrative within science communication and cultural interpretation. The most obvious manifestation of this status quo is the epistemic erasure of non-Western knowledge contributions to science communication. The chapter describes the thorny issues inherent in the context of the production of science communication within this normalised framework of privileges and constructed supremacy. Using case studies from the United Kingdom, France, Canada, the Netherlands, the United States, and from popular culture, and combining the concepts of innocence

(Wekker, 2016) and cultural archives (Said, 1993), this chapter analyses how the legacy of colonialism has entrenched and normalised Eurocentric cultural imperialism in all facets of knowledge communication, including science communication. The chapter also examines the mechanisms of knowledge dissemination with regard to race and colonialism and signposts a transformative agenda for the decolonisation of museums, science centres, and science communication that implies an actionable fall from this normalcy of the Eurocentric normative standpoint across the field.

There is a wealth of learnings, challenges, transformations, and vista-shifting understandings of the decolonisation agenda in science communication that the chapters here bring to this theme. A particularly illuminating insight is brought to bear on the 'umbilical cord' that links the decolonisation of scientific research and the decolonisation of science communication in the Global South regions of the world.

Chapter 11 by Akanimo Odon, highlighting the African scenario, and Chapter 12 by Tibisay Sankatsing Nava, Roxanne-Liana Francisca, Krista T. Oplaat, and Tadzio Bervoets, on the ABCSSS Caribbean Islands, articulate the mechanisms of inequitable North–South research partnerships based on hegemonic Global North funders, which dominate the research agendas/frameworks in Global South regions (see Madise, 2020; Owings, 2021). These 'neo-colonial' transnational research partnerships, described by some as 'helicopter research' (SciDev.net Q&A, 2021), end up creating an underlying pressure on practitioners in the Global South to communicate science through approaches and delivery frameworks that mimic those of their northern counterparts, as determined by international funders. Thus, 'helicopter' science research inexorably leads to 'helicopter' science communication – a double whammy of imposed Eurocentric Global North agendas on the Global South. The unequal power asymmetries at play here are well illustrated by this quote from Carbonnier and Kontinen: 'The very notion of North–South partnerships has turned into yet another development buzzword. Virtually everyone seems to agree with it in principle, but actual practice shows that implementing equitable partnerships is difficult: money flows tend to determine decision-making and actual division of labour' (Carbonnier and Kontinen, 2014).

These two chapters, 11 and 12, open up a crucial and timely conversation that makes the case for the coupling of the decolonisation of scientific research and science communication agendas in order for practitioners in the Global South to inclusively foreground their local challenges, contexts, opportunities, and strengths and thus transform the developmental opportunities and benefits of scientific endeavour and its communication for their local populations. The chapters suggest that one cannot separate how scientific knowledge is produced and generated from how it is communicated – meaning scientific knowledge cannot be inclusively communicated when it has not *also* been inclusively generated, researched, and advanced in the first place.

Part IV: The Globally Diverse History of Science Communication: Deconstructing Notions of Science Communication as a Modern Western Enterprise

Histories of science communication and its Western nodal points have served to render invisible the very long and globally diverse historical footprints of the science communication enterprise in many parts of the world, particularly so in many regions of the Global South. Science communication histories should extend their reach far beyond the Global North to redress those consequences and cultivate a more culturally inclusive ethos and narrative within science communication. This is particularly pressing given that myriad examples abound of the longstanding traditions, practices, and Indigenous knowledge dissemination of many communities across the world, which for centuries have used their own uniquely localised contexts and approaches for the communication of their knowledge and innovation assets. This is an aspect of the inclusion challenge that the field is just beginning to get round to engaging with in terms of how the historical marginalisation of the contributions of non-Western regions to science communication has contributed to the current 'normalisation' of the exclusion of these regions of the world from the science communication agenda (Rasekoala and Orthia, 2020; Orthia et al, 2021).

Inherent in these discourses is the critical appraisal of the multiple platforms, practices, and scenarios in science communication through which these omissions can be redressed. Thus, for example, how do Western museums include the scientific knowledge and concepts of today's diverse communities' ancestors into their communication of science in an inclusive way? Rasekoala and Orthia (2020) argue that this major 'blind spot' in science communication's historical landscape amounts to nothing less than a wholesale repudiation of the contributions of much of the peoples of colour across the world and can ultimately be deemed as racist.

The chapters on this theme demonstrate, feature, and elaborate critical exemplars of some of the trans-cultural examples of these longstanding science communication histories and their practices, epistemic traditions, and innovations from diverse parts of the world. Thus, Chapter 13, by Ruoyu Duan, Biaowen Huang, and Lindy A. Orthia on ancient China, makes the pertinent argument that science communication practices in Western contexts can benefit from the active incorporation of non-Western examples, in this case that of Shen Kua's 11th-century work *Meng Hsi Pi T'an*. This is the earliest notebook encyclopaedia to have been produced in ancient China and offers much of value to history, science, culture, society, and science communication. The origins of this chapter also shed much light on its pertinence to this theme. The chapter was inspired by an Australian postgraduate class in science communication history during which the lead author, an international student from China, had to 'campaign' in order to focus her main research assignment on this text, 'enabling her to claim space within the discipline for her own culture and heritage'. This was due to the very Western emphasis of the course literature, materials, and resources.

The need to challenge such Western dominance in the literature of science communication courses globally is a clear takeaway from this chapter. The experience of the lead author is one that will resonate with many students as well as the academics supervising/teaching the myriad advanced undergraduate and postgraduate science communication courses in many countries. These challenges highlight the profound need to provide critical, diverse, and globally relevant and inclusive teaching, learning, and research resources to support the growing landscape of the many science communication/research/public engagement teaching, training, and capacity-building courses/modules across the globe.

The nature of historical and contemporary migration and the diverse diasporas across the globe (both North and South) also provide an interesting dimension and rationale to this theme. It brings to mind considerations of the potential benefits for diaspora, refugees, and displaced-people communities in particular through the inclusion of the historical, philosophical, cultural, and communication modes embedded in their ancestry in science communication contexts. It also pinpoints the efficacy of this inclusion for transformative science communication in the localities where they are currently based.

In Chapter 14, Siddharth Kankaria, Anwesha Chakraborty, and Argha Manna describe the rise of the printing enterprise and its leveraging in colonial-era Bengal, in India, of the Bengali language and locally contextualised visual illustrations in publications. These developments during this period in India are a direct affirmation of the role of 'bottom-up' knowledge brokering and dissemination approaches in challenging unequal power asymmetries and sociocultural hegemonies within Indian society. They further illustrate the 'people power' concept of inclusive science communication and its impact on the democratisation of knowledge through breaking down socially constructed barriers of race, social class, and caste. The chapter includes lessons for the contemporary era based on the transferrable good practices of this period in India. These lessons signpost and illustrate key features such as the foregrounding of bottom-up participatory knowledge development and dissemination practices, rather than locating them at the margins of science communication, in order to achieve their 'mainstreaming'.

Conclusion

By signposting and describing in detail the deeply challenging yet transformative journey that some science communication institutions and practitioners have undertaken across diverse regions and countries of the globe, some for over ten years, this edited collection seeks to showcase role models for institutional, individual practitioner, and systematic change. The book chapters demonstrate that the road to innovative transformation and decolonisation in science communication is a challenging but rewarding one. In this regard, this is a book that might make for uncomfortable reading for some, but in a sense that is part of its objective – to shift science communication practices in the Global North and South from their complacent comfort zones, blind spots, 'business as usual', and superficial one-dimensional understandings of diversity, equity, and inclusion in the field.

Each chapter has been written with the diverse range of stakeholders in the science communication field in mind – practitioners, academics, policy makers, researchers, institutions, science centres and museums, civil society organisations, and scholars/researchers undertaking undergraduate and advanced science communication training, research courses/modules. The themes, narratives, and signposting to resources have been inclusively curated to ensure that each of these readerships can relate to and learn something about themselves, where they fit, how they are part of the problem of race and sociocultural exclusion in science communication, and, therefore, how they can be part of the solution to this problem going forward. To this end, this book has thus elaborated the shared burden that science communicators in both the Global North and South bear with regard to the challenges and opportunities for race and sociocultural inclusion in the field across these diverse regions of the world.

References

Asante, M.K. (2007) *An Afrocentric Manifesto: Toward an African Renaissance*, Cambridge: Polity Press.

Carbonnier, G. and Kontinen, T. (2014) 'North–South research partnerships: academia meets development?, EADI Policy Paper Series, Bonn.

Dawson, E. (2019) *Equity, Exclusion and Everyday Science Learning: The Experiences of Minoritised Groups*, London: Routledge.

Finlay, S.M., Raman, S., Rasekoala, E., Mignan, V., Dawson, E., Neeley, L. et al (2021) 'From the margins to the mainstream: deconstructing science communication as a white Western paradigm', *JCOM: Journal of Science Communication*, 20(1): CO2. doi: https://doi.org/10.22323/2.20010302

Gerber, A. (2018) 'Panel session abstract: "Opportunities and risks of a globalisation in science communication; future perspectives for the PCST network"', private communication/invitation to panel session proposal for the PCST Conference held in Dunedin, New Zealand, in 2018.

Hikuroa, D. (2017) 'Matauranga Maori: the ukaipo of knowledge in New Zealand', *Journal of the Royal Society of New Zealand*, 47: 5–10.

Kago, G. and Cissé, M. (2022) 'Using African Indigenous languages in science engagement to increase science trust', *Frontiers in Communication*, 6, Available from: https://www.frontiersin.org/articles/10.3389/fcomm.2021.759069/full

Lewenstein, B. (2019) 'The need for feminist approaches to science communication', *JCOM: Journal of Science Communication*, 18(4): C01. doi: https://doi.org/10.22323/2.18040301

Madise, N. (2020) 'Level the playing field for science in the global south', SciDev.net, Available from: https://www.scidev.net/global/cooperation/opinion/level-the-playing-field-for-science-in-the-global-south.html

Márquez, M.C. and Porras, A.M. (2020) 'Science communication in multiple languages is critical to its effectiveness', *Frontiers in Communication*, 5 (May). doi: https://doi.org/10.3389/fcomm.2020.00031

Olatunji, F.O. and Bature, A.I. (2019) 'The inadequacy of post-development theory to the discourse of development and social order in the Global South', *Social Evolution and History*, 18(2) (September). doi: https://doi.org/10.30884/seh/2019.02.12

Orthia, L., Hikuroa, D.C.H., Nabavi, E., Rochberg, F. and De Vos, P. (2021) '3 reasons to study science communication beyond the West', The Conversation, 12 January, Available from: https://theconversation.com/3-reasons-to-study-science-communication-beyond-the-west-152237

Owings, L. (2021) 'Research colonialism still plagues Africa', SciDev.net, Available from: https://www.scidev.net/sub-saharan-africa/scidev-net-investigates/research-colonialism-still-plagues-africa/

Potter, N. (2018) 'What is Eurocentrism? Locating a suitable definition for qualitative research on the possible Eurocentric bias in International Relations (IR)', LSE Impact of Social Sciences Blog, 18 April, Available from: https://blogs.lse.ac.uk/internationalrelations/2018/04/18/what-is-eurocentrism-locating-a-suitable-definition-for-qualitative-research-on-the-possible-eurocentric-bias-in-ir/

Q&A (2021) 'How to stop "helicopter research colonisation"', SciDev.net, Available from: https://www.scidev.net/global/opinions/qa-how-to-stop-helicopter-research-colonisation/

Quijano, A. (2000) 'Coloniality of power, Eurocentrism, and Latin America', *Nepantla: Views from South*, 1(3): 533–80.

Rasekoala, E. (2016) 'Truly socio-culturally inclusive, "colourful" and gender-balanced', Blog in Ecsite Spokes Online Magazine, Available from: https://www.ecsite.eu/activities-and-services/news-and-publications/digital-spokes/issue-20#section=section-column&href=/feature/column/truly-socio-culturally-inclusive-colourful-and-gender-balanced

Rasekoala, E. (2020) 'Challenges and opportunities for transformation', in Ecsite webinar 'Decolonising museums and science communication', 13 May, Available from: https://www.youtube.com/watch?v=CWxw0x7U5lw

Rasekoala, E. and African Gong (2016) 'ISCSMD 2016: a global platform for reflection, transformation and mainstreaming the SDG's in science centres and museums', 11 July, Available from: http://www.iscsmd.org/2016/07/11/iscsmd-2016-a-global-platform-for-reflection-transformation-and-mainstreaming-the-sdgs-in-science-centres-and-science-museums/

Rasekoala, E. and Orthia, L.A. (2020) 'Anti-racist science communication starts with recognising its globally diverse historical footprint', LSE Impact of Social Sciences Blog, 1 July, Available from: https://blogs.lse.ac.uk/impactofsocialsciences/2020/07/01/anti-racist-science-communication-starts-with-recognising-its-globally-diverse-historical-footprint/

Said, E. (1978) *Orientalism*, New York: Pantheon Books.

Said, E. (1993) *Culture and Imperialism*, Vintage digital.

Seleti, Y. (2012) 'The value of Indigenous knowledge systems in the 21st century', in J.K. Gilbert and S.M. Stocklmayer (eds) *Communication and Engagement with Science and Technology*, New York: Routledge, pp 267–78.

Wade, L. (2019) 'Decolonizing science communication by telling regional stories', *American Journal of Physical Anthropology*, 168: 259–60.

Wekker, G. (2016) *White Innocence, Paradoxes of Colonialism and Race*, Durham, NC: Duke University Press.

Woldeyes, Y.G. (2019) 'Repatriation: why Western museums should return African artefacts', The Conversation, 15 May, Available from: https://theconversation.com/repatriation-why-western-museums-should-return-african-artefacts-117061

PART I

The Practice(s) of Science Communication: Challenges and Opportunities for Race, Gender, Language and Epistemic Diversity, Representation, and Inclusion

1

Inclusion Is More Than an Invitation: Shifting Science Communication in a Science Museum

C. James Liu, Priya Mohabir, and Dorothy Bennett

Introduction

> I'm from the Bronx and I travel two hours away to Queens to teach science because it makes me happy. And then seeing the faces of other little kids of colour when they like, see me do a 'Cow Eye Dissection'. They might have never seen a Black person do that before. And they might have never seen a Black person as a scientist before. It's amazing because they're being exposed to an experience that they never thought they would ever have had. (Explainer E [17 years old])

As popular, public-facing institutions that are often perceived to be authoritative and trustworthy, science museums play an important role in public science education (National Research Council [NRC], 2009; Falk et al, 2012; National Science Foundation [NSF], 2012; Falk and Dierking, 2018; American Alliance of Museums [AAM], 2021). Through engaging with visitors and local communities, science museums contribute to long-term impacts on the public's scientific literacy and understanding (Falk and Needham, 2011) and provide resources for schools and teachers to support students' learning and motivation in science, technology, engineering, and mathematics (STEM) (NRC, 2009).

At the same time, girls, African Americans, Latinx, and Indigenous students of all genders are often marginalised in formal STEM educational settings starting from an early age, are chronically underrepresented in STEM career pathways, and have been consistently shut out of high-quality STEM education and opportunities (Committee on Underrepresented Groups, 2011; McGee, 2020). This is a critical issue because, as has been evident during the COVID-19 pandemic, science literacy and thinking skills are essential to the quality of our decision-making. As science museums, it is our mission to ensure that all children and students have equal access and opportunity to engage with STEM tools and methods to be active participants who create their own world and future and to contribute to the future of STEM professions.

Recently, researchers, educators, leaders, and policy makers in informal science education have raised their voices, demanding inclusiveness, diversity, and equity in the broad ecosystem of STEM education (Dawson, 2014; Feinstein and Meshoulam, 2014; AAM, 2021; Ba et al, 2021; National Academies of Sciences, Engineering, and Medicine [NASEM], 2021). The intention is to surface opportunities that should be available to minority communities and ensure 'children of all backgrounds, race, ethnicity, gender, religion and income levels can learn the wonders and possibilities of STEM and maintain that interest and passion throughout their lives' (NASEM, 2021, p 8). A growing number of initiatives have been taking place to address the need for more inclusive informal science education. For example, programme developers are increasingly embracing asset-based and culturally responsive approaches to inviting families from diverse backgrounds to bring in their culture and build personal connections into their museum visits (see Zollinger and DiCindio, 2021). Researchers are also encouraging a reconceptualisation of evaluation and assessment in informal education to better understand marginalised children's and their families' experiences in these environments (Garibay and Teasdale, 2019). The message is clear: inclusion, equity, and diversity are the mission of science museums and other informal science education institutions today.

One way to address this issue is by changing the way science communication is conducted in science museums. A body of research, for example, has been investigating families' conversations while engaging with exhibits at science centres and how such conversations shape their experiences (Leinhardt et al, 2003; Haden et al, 2014). Recent studies on inclusive science communication have also indicated the power and value of moving away from merely communicating scientific concepts to making the connection that science is a practice we all take part in and have experienced in different ways (Canfield et al, 2020; Ba et al, 2021; Letourneau et al, 2021). At the New York Hall of Science (NYSCI), science communication is part of a participatory experience that invites visitors to explore, ask questions, gain skills, and solve problems, taking multiple pathways towards understanding concepts and phenomena that connect back to their culture, perspectives, and daily experiences.

We use the art of dialogues, back and forth between facilitators and visitors, to create an environment in which people are encouraged to share stories and make connections, and use the space we created, including exhibits, signage, activities, and programmes, to highlight the essential role that science communication plays in defining, practising, and learning science. Rethinking and reimagining what science communication is about, looks like, and for whom can help science museums create engaging, welcoming, and inclusive science education for all.

In this chapter, we share how NYSCI tackles inclusion, equity, and diversity through a shift in our approach to redefining science learning and particularly science communication. We first introduce our pedagogical approach and youth museum educators, known as Explainers, and then provide reflections on some of the important shifts in practices that took place from the perspectives of Explainers, trainers, and the director of the youth development programme. The reflections illustrate the different risks, perspectives, and roles that people take while communicating science, and how their identity, culture, and race influence the choices they make when engaging with different audiences.

Our approach

Located in Queens, New York, NYSCI serves one of the most diverse counties in the United States (Lobo and Salvo, 2013). The museum features a wide range of STEM topics, from the classic 'Bubble' exhibit, where children use iron hoops to make giant bubbles, to 'Design Lab' – a 10,000-square-foot exhibition space entirely dedicated to hands-on, engineering design activities. The exhibits support STEM learning through an approach called 'Design, Make, Play' (Honey and Kanter, 2013) that emphasises learning through making personal connections, problem-solving, and playful exploration. The approach is guided by developmental and cognitive psychology and is rooted in constructivism and sociocultural theory (Lave and Wenger, 1991). It frames learning as a social practice co-constructed by groups of learners and positions people at the centre of creating meaningful learning experiences. By adopting this approach, our goal is to *invite* and *be invited* into a process of sharing and connecting everyone's perspectives and experiences with STEM.

The approach has developed gradually over the last ten years as NYSCI has reimagined our visitor experience. We have moved away from focusing on offering visitors engaging explanations of scientific concepts and phenomena towards more participatory experiences that emphasise visitors' agency. One key to bringing about this shift is transforming the roles of our young museum floor educators, Explainers, and redefining the ways we think about science communication. In practice, Explainers have to shift their focus from explaining the science content and concepts behind exhibits to supporting visitors in following their own questions and designing their own solutions to problems. At the same time, they also have to reflect on the ways visitors engage with STEM to help shape what these interactions can look like. In the end, it is this reflection that informs our practices, creating a dynamic cycle for improving visitor engagement.

Who are the Explainers?

When we think about our societal structure and who sits in a place to help us learn and explore the world around us, we typically think about adults, whose academic training and life experiences have equipped them with the authority and ability to share their knowledge with others. But who says that we can't learn from and with the young members of our communities, the individuals that are typically seen as the learners? At NYSCI, it is our Explainers (ages 14–25), participants in NYSCI's signature Science Career Ladder (SCL) programme, who are positioned on the museum floor to engage with visitors at our exhibits, programmes, and events. They are the ones who help bring the ideas of science to life and help visitors explore the process of doing science.

To say that our Explainers offer diversity to the museum floor is an understatement. The Explainers represent a wide range of ethnic, racial, and socio-economic backgrounds and have diverse lived experiences from being first-generation college students, many of whom have recently migrated to the United States to work towards a better future for themselves and their families. We have Explainers who knew they loved science at a young age and use this experience to deepen their understanding and broaden their experiences, and others who stumbled on the opportunity and are developing an appreciation and understanding of science. No matter their backgrounds or paths, the Explainers are active participants in a community of learning that fosters their own interest and comfort with exploring questions. In turn, their curiosity enriches the kinds of interactions they have with visitors. Explainers are paid part-time employees, which encourages students from underrepresented backgrounds to participate more fully in these programmes, especially if they need to financially contribute to or support their families.

Shifting science communication by reflecting on personal relationships with science

In the process of shifting science communication on our museum floor, one thing that really stands out is our Explainers' and trainers' personal relationships with science. When we tried to find a new way to connect with visitors, we found that we cannot just ask visitors to openly discuss their experiences with us but had to put ourselves in the conversation and share our thoughts and experiences. We noticed that conversations often started from our personal stories about encountering scientific concepts or prior relevant experiences. As we tried to construct conversations that invite in visitors' culture, backgrounds, and stories, we naturally brought in our own assumptions, values, and beliefs about science shaped by years of formal and informal learning experiences with science and people. Through this process, we realised that diversity and inclusion start from seeing and knowing not only our visitors but also ourselves.

This connection to our personal relationships with science echoes the asset-based and community of practice approaches to learning that acknowledge how our journeys into science can vary greatly according to the roles we choose or are assigned by others, how we present ourselves, and how we are seen and treated by others (Adams and Gupta, 2013). Through reflections, we uncovered the different histories and personal relationships Explainers and staff had with science and how these differences may contribute to their personal beliefs and practices in science learning and communication. Some gained support and inspiration from others who strengthened their interests and identification with science, while others experienced science learning as difficult and distant, not being able to identify themselves with science or build competence as science learners or educators:

> My dream was to become a paediatrician, and I held on to it for a LONG time. I took steps to integrate science into my life. I attended Queens High School for the Sciences at York College with the intent of taking more science courses and having easy access to the Sophie Davis Health Professionals Mentorship Program. I also began as a volunteer at NYSCI. The hope was that these two institutions could nurture me right into med school. … Things did *NOT* go as planned. My lowest marks at school were always my science courses, and even at NYSCI, I struggled to connect with the content required to excel in the training model at the time. By my sophomore year in high school, I'd convinced myself that science was not my forte, and that I never liked it in the first place. I spent the next two years in my science high school and with my science job feeling like a failure. (Trainer A)

> My high school earth science teacher was a Black Caribbean man that for some reason thought I was a brilliant student. He encouraged me to keep pursuing science and told the chemistry teacher to expect greatness from me in the next grade level. … [However], the chemistry teacher had a different style of teaching than that of the earth science teacher. Unfortunately, I didn't do as well in chemistry, and the teacher showed how unimpressed he was

> by my underwhelming performance in his class … [but] I did well on standardised tests and that got the attention of my high school principal who noticed my interest in science and encouraged me to read the *Science Times* on Tuesdays. They didn't sell the *Times* in my neighbourhood in Brooklyn back then, so she took it upon herself to bring me a copy of the *Science Times* every Tuesday. But not until I started my time at NYSCI did I really find a way to connect with all of the content I took in. (Trainer T)

The situation can be even more challenging for first-generation immigrant children and families in the United States, who need to navigate language barriers and cultural differences:

> Most of the educational support I had grown accustomed to was demolished when mom, two of my siblings, and I migrated to New York and settled in Brooklyn. Our new life meant my mother now worked to provide. I was now without my oldest sister who served the role of an authoritative sibling and teacher. The culture shock made transitioning difficult – I went from being supported by teachers and peers to being almost ridiculed for using the letter 'u' when spelling colour or neighbour. There were stark differences in the manners of teaching and the ways the teacher engaged with the students. Rather than ask questions in maths or English class, I called on some of the lessons I learned in Guyana only to be told: 'that's right, but that's not how we do things here'. I wish I had the gall like some of my classmates to say the way you do things doesn't work for me – but that wasn't the culture I was raised up in. Undoubtedly, I am sure my accent and the teacher's responses caused me to become less vocal in the classroom. (Trainer I)

These personal experiences are powerful. They influence how we think and believe science should be learned and communicated and how we might think about and approach equity and inclusion. For example, the challenge that Trainer A experienced had a direct impact on the way she creates the learning environment for visitors and Explainers:

> My passion for education and nurturing young learners stems directly from my own struggle to connect with material that everyone else around me seemed to be grasping. It is a feeling I don't wish on anyone, so it more than influences my teaching style. … When leading training or even just having a conversation, I often pay attention to the room, scanning facial expressions and body language for clues that someone might feel left out. I'll often present an idea in a few different ways

> (for example, verbally, visually, and physically) until most heads are nodding, or even ask willing volunteers to share how they connect with a topic, in hopes that one more person can feel included. (Trainer A)

On the other hand, Trainer T reflects his experience of learning science with others in his training style. He focuses on creating an open space in which everyone is a teacher and learner:

> As a Trainer, I've used aspects of my family experience, my academic experience, and my experience at NYSCI to influence my training style. One of my training goals is to always be learning. I learn from the Explainers and other Trainers. As a kid, it always felt like the adults in my life only wanted to impart knowledge to me but didn't ever see knowledge as a mutual exchange. It's important that I learn from everyone around me. The Explainers who claim to be 'not science' people have often taught me some of my most memorable lessons – how to think differently about a topic. (Trainer T)

Personal reflection is also important for Explainers to find their own ways to connect with science and share that with the visitors in their work:

> [L]ike in high school, I was a very … like theatre dance, you know. I was a little bit far from science-y stuff. But what drew me to the museum was, I guess, inclusive and made science … very welcoming in terms of like, 'oh, science is actually something that's possible and something that could be really easy'. And Explainers really helped in that mission. And I really wanted to see myself as a part of it. (Explainer J)

For visitors, this approach translates into a shared space where personal connections and learners' agency are encouraged. This can be a unique opportunity for families, especially families who had limited museum experience, to see a different way of learning, as one parent and child reflected:

> I think that from my experience, I think it's a time that should be taken out NOT specifically for telling [our children], 'ok, sit, I'm going to show you this', but also to see that they have a lot of creativity and that they can also teach us. Not only us, but a lot of times they design and through play, they can teach us a lot of things that we forgot … it's that, spending some time with them. (Parent S)

> That everyone got to interact with each other, to learn new things about each other. We got to learn science in a different way than the rest. Like in schools we usually just write them on the board and

> sometimes they don't explain it thoroughly with a project or something like that but here you can actually see everything in action. (Child K)

It was clear to us that inclusive science communication is more than an invitation. We learned that once we took the very first step of expanding our focus beyond relaying scientific facts about phenomena, conversations with visitors and within our staff became relatable, passionate, and genuine. Our Explainers are able to bring themselves into the conversations and make connections with people. Perspectives are then not artificially invited but organically generated and included across members of the conversations. In this environment, personal experiences with science were celebrated and cannot be isolated from science communication. Offering opportunities for our Explainers to bring their own personal histories with science then allowed them to reconsider that initiating inclusive and engaging science communications doesn't depend on expertise but on experiences and openness. This approach encouraged them to follow visitors' leads, inspire their curiosity and questions, and support visitors in pursuing their own distinctive approaches to problems.

The impact of Explainer experience

When everyone is bringing their own experience into our museum, when we listen to everyone's voices, how can we ensure that science communications are engaging and educational for a diversity of visitors and the public? This challenge is addressed in the SCL programme, which establishes common principles and practices for our Explainers. In fact, most of our staff in the Explainer programme were once Explainers themselves. This connection establishes two important foundations for the Explainer programme. Firstly, the Explainer programme provides an opportunity for our SCL staff and Explainers to connect with science through our Design, Make, Play approach that focuses on multimodal learning through hands-on explorations of phenomena and relatable problems identified by a diverse team of staff and Explainers. These shared experiences open up a unique way of learning science that is often new to many of our staff and Explainers:

> Engaging with science at NYSCI really changed the way I thought about learning. I always enjoyed reading and watching science, and for most of my life that was the way I thought I learned best. In high school, I knew if I read the textbook, I would understand the concepts well enough to pass the exams. NYSCI changed that because I got to engage with so many things. I learned with all of my senses. Those interactions really made things that I thought I understood really make sense. (Trainer T)

Secondly, this shared experience helps SCL staff to continue to modify and develop Explainer training through their own reflections on the ways these approaches might engage visitors with different science learning histories and identities. These experiences resulted in shifting the goals of Explainer training away from just knowing the content to providing opportunities for visitors to engage with science practices. Trainers focused on developing the Explainers' ability to confidently investigate exhibits, concepts, and ideas alongside the visitors even when they are unsure of the outcome. Through the development of these strategies, the goal was to build their capacity to welcome different perspectives and create conversations centred on what the visitors find most interesting or relevant and engage with them based on their own personal experiences and perspectives. Science content shifted from being the end goal of these conversations to a resource to open up the questions in these interactions. The shifts in the ways Explainers were trained also facilitated shifts in how the Explainers thought about science and our role as a museum:

> I feel like [the museum] is for everyone … just grabbing different people from different countries and different cities and just showing them what is science, you know, how can science be applied in your everyday life? Science is not just in the textbooks, but you can physically do it in cool and imaginative ways. (Explainer L)

> The most useful [part of our training] is where you get to listen to everyone's ideas and no one's idea is shut down. When we say, "this is what I do" and like, "this is what someone else does", and that's perfectly fine. And everyone has different beliefs of what they should do. That actually has helped me where I get to look at other people's point of view and see how I could better myself. (Explainer D)

The experience of being an Explainer has also allowed our training team to develop confidence in creating experiences for this generation of Explainers that builds their agency. Through our roles, we have learned the importance of flexibility and have created ways for Explainers to make their own choices that reflect both their voice and experiences in engaging visitors. We have learned that the Explainer role is about allowing both the Explainers and visitors the opportunity to connect with science in ways that are meaningful to them and that build on their diverse experiences and interactions with STEM:

> Having an understanding that everyone may have different interests and expertise has made it clear that there is no cookie-cutter example when it comes to engaging and training our Explainers. …

> The experiences I've had growing up in conjunction with my year at NYSCI seem to have come full circle. The tinkering I did as a child while I did not see it as engineering or involving any aspect of STEM continues to find its way into my training and leisure activities. (Trainer I)

Overall, the shared experience of being an Explainer helped the SCL team understand the impact and potential of the Explainers' work and the importance of offering opportunities for Explainers to think critically, reflect, and build on their own experiences with science as a way to adapt an exploratory approach to communication that sets up a more inclusive learning environment that can potentially draw in many more learners, as summarised in the following excerpt by the SCL programme director, who was also once an Explainer:

> When I first entered NYSCI and put on the red apron it was intimidating to think that I was going to be teaching people science. I was 17, my teachers had gone through years of education and experience in order to teach me, and here I was trying to do that before I even finished high school! That's what it was at the beginning – a feeling that I had to teach people new things about the exhibits and in turn the world around us. I didn't know what I didn't know, and I didn't recognise the power of having someone like me, a 17-year-old Brown girl, born in Guyana, with a Muslim dad and Hindu mom, a sister, a cousin, and a hidden nerd, on the museum floor sharing information and engaging visitors. It's taken many years for me to see that for myself and I use that so that our current Explainers have a better understanding of their value and importance.
>
> I have gone through learning experiences to help shape the ways I engage with visitors. I have been pushed outside the boundaries of my own comfort to help develop not just a voice, but my voice in the ways I shaped that engagement. I have felt why it's important to have people like me be the ones engaging our visitors, primarily families with children, across the museum floor and beyond. I have had to question my own understanding and relationship with STEM, undo some of my own training in order to develop new ways of working with youth and inviting them to contribute to the ways we engage others in STEM. I have reflected on my experiences and discovered why they matter, why I matter, and how they have helped shape the direction we have taken over the last six years. I know what it is that we are asking of our Explainers and how uncomfortable that can sometimes be. (Programme Director P)

Tools for reflection and structural changes to support a shift towards more inclusive communication practices

This shift in how Explainers invite science communication at NYSCI was not easy to start and is still evolving and developing. It took a decade to explore, learn, and reflect on our approach and practices. In the following, we summarise key strategies developed from different projects that contributed to the shift over the years, all with the goal of elevating both the Explainers' and visitors' interests, questions, and experiences.

A framework for visitor-centred engagement – Explainer habits of mind

At the beginning of our shift, although not clearly articulated, we felt the urgency of finding a shared language to describe the kind of science engagement and communication that we are aiming for. Drawing on several prior works (for example, NASEM, 2018) and iterations across multiple projects (see Mohabir et al, 2021), we developed a set of Explainer habits of mind (including curiosity, engagement, persistence, flexibility, empathy, and being personable) for our work with Explainers. These habits of mind help Explainers respond nimbly to the diverse needs of visitors with different perspectives, cultural backgrounds, experiences, and interests. For example, in considering the habit of *being personable* in her work on the floor, one Explainer shared how this is "a huge skill to have to kind of get over that hurdle of just initiating conversations with people" even if she was new to the space or exhibit. While some found adopting the habits of mind challenging, many Explainers described how thinking about the habits of mind as something that everyone uses in different ways helped them ask critical questions, bring new skills into their practices on the floor, and improve their work with the help of supportive peers and staff working towards the same goals. These mind-sets have become vital for fostering conversations with visitors and encouraging Explainers to reflect on their own growth and the growth of their peers:

> I think the idea of the Explainer habits of mind impacted our work by really thinking about what are the things that we want Explainers to be able to achieve and how would we want them to succeed as Explainers. It's not just about the Explainer growth, but that Explainer growth also helps our visitors to grow and to feel confident and to want to try new things. (Exhibit Designer S)

A cross-departmental collaboration

Rather than relying only on the voices of staff who were responsible for training the Explainers, an additional benefit of developing the Explainer habits of mind was that it established a cross-departmental collaboration that contributed to an institution-wide shift in leadership style (for details, see Ba et al, 2021; Letourneau et al, 2021). The team included Explainers, trainers, researchers, programme developers, and exhibit designers. The Explainers offered their first-hand experiences interacting with visitors; the trainers articulated the core needs and practices that came into play in youth development; the researchers provided in-depth knowledge about child-centred approaches to STEM teaching and learning; programme staff brought up the needed resources and operational concerns for public programming; and the exhibit designers illustrated the intention behind museum exhibits and activities. These diverse perspectives, and the time we devoted to integrating these voices, were critical to enriching the work of supporting Explainers and, more importantly, to lay a foundation for widespread institutional change. This cross-departmental approach ensured that work was not 'owned' by one department but, instead, was better defined and recognised through the practices we were trying to develop. By integrating Explainers' perspectives and experiences, the team was able to develop a shared language about the factors that influenced visitors' learning and the challenges that sometimes arose in implementing new facilitation strategies. The shift in authority and sharing responsibility has resulted in great success in improving our practices in addition to the development of the Explainer habits of mind and has become the standard practice in almost every aspect of our work – to bring in Explainers' voices, experiences, and ideas.

Co-designing demonstrations and programmes

Explainers have also been invited to play an active role in co-designing facilitated experiences and training materials to ensure that not only were their voices heard and valued but that they can also contribute directly to the design of visitor experiences. This is evident in the development of a series of new science demonstrations and programmes at the museum.

Traditionally, the demonstrations usually consisted of Explainers illustrating a scientific principle by displaying how something works, using a scripted line of actions and questions that highlight the concepts involved (demonstrating the properties of matter by using liquid nitrogen, for example). Specific scripts were typically created by our full-time staff (scientists and training team) and then brought to the Explainers as a way to train them on how to conduct a variety of STEM-based demonstrations to ensure that the science facts and information shared were accurate, and to build Explainers' confidence with content. Although there was always an invitation to modify these demonstrations to better engage diverse audiences, that invitation never translated into practice, and for a long time we did not understand why. This changed as we reflected on the evolving nature of the Explainers' role and began to present demonstrations as a series of activities that could be reconfigured to tell a story in the way each Explainer wanted to tell it, offering suggestions for themes in demonstration guides rather than scripts. By reformatting our training in this way, we started to see Explainers creating their own narratives that are exciting to them and that allow them to bring their experiences and viewpoint into their demonstrations. This approach has inspired a continuing reworking of our other traditional, longstanding demonstrations, such as 'Cow Eye Dissection', to allow Explainers to have more of a voice and an active role in deciding the direction and outcome of the demonstrations.

Additionally, in our more recent practice, we have asked the Explainers to be the drivers of programme activities and have helped support them in bringing their own cultural perspectives into creating activities that show how scientific principles are part of their lives. For example, one group of Explainers was interested in creating an activity that highlighted the process of fermentation and did so through making kimchi and kefir. Their activity shared this process through their personal experiences with this food and drink, using things they have in their everyday lives, and created an invitation for visitors to explore this science through the lens of their own culture and experiences as well.

Learner-centred pedagogical approach and evaluation

It is important to note that the work with Explainers and the shift in our science practices and communication is guided by the reflective, learner-centred approach in which visitors own the agency of constructing their learning experiences. The underlying principles of this approach, as Ba and colleagues (2021) suggest, include: (a) *putting learners at the centre*, leveraging learners' natural instincts to engage playfully with compelling ideas and materials; (b) *positioning learners as creators*, making it possible for them to actively create their own learning experiences and tackle problems that they think are worth solving, rather than being consumers of content, materials, or material objects; and (c) *cultivating perspective-taking and divergent solutions*, which helps learners as well as educators to build connections to their own experiences, knowledge, and skills, and as a result, providing an accessible invitation for learners from different backgrounds to share their ideas with others and appreciate the diversity of what others think, make, and share. In addition, we are currently developing systematic ways to define and document the impact of this approach from the visitors' point of view. Several preliminary efforts have been made through different projects, including projects funded by the National Science Foundation, to better understand, for example, visitors' intentions and motivation and sense of agency. The goal is to establish foundational evidence that will support field-wide strategies for more inclusive and diverse science communication.

Conclusion

Through the course of our work, we have realised that science communication is not only about our interactions with visitors. It starts with the internal conversations we have among ourselves, the ways we identify science in our everyday lives, and the ways that we invite others into interactions that welcome and celebrate our diverse experiences. We have also found it critical to reflect on our own experiences and interactions with STEM, and to recall how that can shape one's identity and experience of science. These lessons align with prior research emphasising that inclusive science communication requires building on asset-based approaches and contextual expertise of communities (culturally responsive design), lay expertise (multiple ways of knowing), shared identity and characteristics (intersectionality), and public participation (co-creation and collaborative design) (see Canfield et al, 2020 for review). We have also learned and leaned into the idea that to create long-term, sustainable shifts in practices that invite and welcome diverse perspectives and voices, we need a shared goal across the institution to open up our design and development processes so that Explainers can actively contribute to redefining our approaches to visitor engagement. It is critical for allowing leadership to anticipate Explainers' needs for continued development and training, and to provide them with greater opportunities to recognise and use their own agency to create unique and personalised learning experiences that resonate with our visitors.

Our work with Explainers and visitors at NYSCI illustrates how scientific knowledge can be shared, explored, challenged, negotiated, and defined in a public science communication system such as a science museum, thereby inviting the STEM community and stakeholders to rethink and question their approach to inclusion and diversity in science education and communication.

References

Adams, J.D. and Gupta, P. (2013) 'I learn more here than I do in school. Honestly, I wouldn't lie about that: creating a space for agency and identity around science', *International Journal of Critical Pedagogy*, 4(2): 87–104.

American Alliance of Museums (2021) 'AAM museum and trust report', Available from: https://www.aam-us.org/wp-content/uploads/2021/09/Museums-and-Trust-2021.pdf

Ba, H., Culp, K.M., and Honey, M. (eds) (2021) *Design, Make, Play for Equity, Inclusion, and Agency: The Evolving Landscape of Creative STEM Learning*, New York: Routledge.

Canfield, K.N., Menezes, S., Matsuda, S.B., Moore, A., Mosley Austin, A.N., Dewsbury, B.M. et al (2020) 'Science communication demands a critical approach that centers inclusion, equity, and intersectionality', *Frontiers in Communication*, 5: 2.

Committee on Underrepresented Groups and the Expansion of the Science and Engineering Workforce Pipeline (2011) 'Expanding underrepresented minority participation: America's science and technology talent at the crossroads', Washington, DC: National Academies Press.

Dawson, E. (2014) 'Equity in informal science education: developing an access and equity framework for science museums and science centers', *Studies in Science Education*, 50(2): 209–47.

Falk, J.H. and Dierking, L.D. (2018) *Learning from Museums*, Lanham, MD: Rowman & Littlefield.

Falk, J.H. and Needham, M.D. (2011) 'Measuring the impact of a science center on its community', *Journal of Research in Science Teaching*, 48(1): 1–12.

Falk, J., Osborne, J., Dierking, L., Dawson, E., Wenger, M., and Wong, B. (2012) 'Analysing the UK science education community: the contribution of informal providers', Wellcome Trust.

Feinstein, N.W. and Meshoulam, D. (2014) 'Science for what public? Addressing equity in American science museums and science centers', *Journal of Research in Science Teaching*, 51(3): 368–94.

Garibay, C. and Teasdale, R.M. (2019) 'Equity and evaluation in informal STEM education', *New Directions for Evaluation*, 161: 87–106.

Haden, C.A., Jant, E.A., Hoffman, P.C., Marcus, M., Geddes, J.R., and Gaskins, S. (2014) 'Supporting family conversations and children's STEM learning in a children's museum', *Early Childhood Research Quarterly*, 29(3): 333–44.

Honey, M. and Kanter D.E. (eds) (2013) *Design, Make, Play: Growing the Next Generation of STEM Innovators*, Abingdon: Routledge.

Lave, J. and Wenger, E. (1991) *Situated Learning: Legitimate Peripheral Participation*, Cambridge: Cambridge University Press.

Leinhardt, G., Crowley, K., and Knutson, K. (eds) (2003) *Learning Conversations in Museums*, New York: Taylor & Francis.

Letourneau, S.M., Bennett, D., McMillan Culp, K., Mohabir, P., Schloss, D., Liu, C.J. et al (2021) 'A shift in authority: applying transformational and distributed leadership models to create inclusive informal STEM learning environments', *Curator: The Museum Journal*, 64(2): 363–82.

Lobo, A.P. and Salvo, J.J. (2013) 'The newest New Yorkers: characteristics of New York's foreign-born population', Population Division of the New York City Department of City Planning, Document #NYC DCP #13–10, Available from: https://www1.nyc.gov/assets/planning/download/pdf/planning-level/nyc-population/nny2013/nny_2013.pdf

McGee, E.O. (2020) 'Interrogating structural racism in STEM higher education', *Educational Researcher*, 49(9): 633–44.

Mohabir, P., Bennett, D., Liu, C.J., and Tetecatl, D. (2021) 'From explaining to engaging visitors: transforming the facilitator's role', in H. Ba, K.M. Culp, and M. Honey (eds) *Design, Make, Play for Equity, Inclusion, and Agency: The Evolving Landscape of Creative STEM Learning*, New York: Routledge, pp 28–43.

National Academies of Sciences, Engineering, and Medicine (2021) *Call to Action for Science Education: Building Opportunity for the Future*, Washington, DC: National Academies Press.

National Academies of Sciences, Engineering, and Medicine (2018) *How People Learn II: Learners, Contexts, and Cultures*, Washington, DC: National Academies Press.

National Research Council (2009) *Learning Science in Informal Environments: People, Places, and Pursuits*, Washington, DC: National Academies Press.

National Science Foundation (2012) 'Science and engineering indicators 2012', Arlington, VA: National Science Foundation.

Zollinger, R. and DiCindio, C. (2021) 'Community ecology: museum education and the digital divide during and after COVID-19', *Journal of Museum Education*, 46(4): 481–92.

2

Communicating Science on, to, and with Racial Minorities during Pandemics

John Noel Viana

Introduction

Pandemics, such as Severe Acute Respiratory Syndrome (SARS) and COVID-19, can disproportionately impact migrants and racial minorities through increased morbidity and mortality (Pan et al, 2020) and limited consideration of their needs in measures to control disease spread (Tan, 2021). Furthermore, communication initiatives to educate the public about SARS and COVID-19, particularly their origin, spread, and control, can lead to stigmatisation, othering, and exclusion of Asian minorities (Hung, 2004). Although communicating the science associated with these infectious diseases can facilitate transparency and rationalise lockdown measures, they can also harm minority ethnic groups and the broader social fabric when conducted in a culturally insensitive and exclusionary manner.

To illustrate the importance of sensitive and inclusive pandemic science communication, this chapter draws from accounts of two coronavirus pandemics. Before the 2020 COVID-19 pandemic, the world was threatened by SARS in 2003, especially the Canadian city of Toronto. Communication on its origin in China led to racially motivated attacks and discrimination against people who look East or South East Asian and against businesses in Toronto's Chinatowns (Keil and Ali, 2006). These forms of discrimination were also experienced by Australia's Asian minority population during the COVID-19 pandemic (Asian Australian Alliance and Chiu, 2020). However, the management of COVID-19 in Australia also highlighted the disproportionate impact of lockdown policies on racial minority and socio-economically disadvantaged groups, especially with inadequate communication of their implementation (Victorian Ombudsman, 2020) and scientific/epidemiological rationale (Patrick, 2021). These were further aggravated by limited engagement with community members in planning lockdowns (Victorian Ombudsman, 2020).

This chapter draws from academic publications, reports, and news articles on SARS and COVID-19 to illustrate how different forms of communication on and during these pandemics profoundly affected the welfare of racial/ethnic minorities. Lessons from these incidents can be used to develop more inclusive ways of communicating pandemic science and formulating associated policies (Hyland-Wood et al, 2021). Experiences during SARS and COVID-19 can help develop pathways not just for communicating science involving racial minorities but also for relaying scientific information that has a profound impact on them. Going beyond communication on and to, lessons during these pandemics are vital in underscoring the importance of engaging with minorities to develop culturally sensitive communication strategies (Airhihenbuwa et al, 2020). These pandemics demonstrate the limitations of a deficit model in science communication – wherein scientists, science communicators, and politicians fill in a perceived knowledge gap of the public (Reincke et al, 2020) – and point to the need for a more engaged model, whereby the lived experiences and voices of racial minorities are valued and taken into account.

SARS communication in Canada

SARS is a coronavirus disease first recorded in animal handlers in Guangdong, China, in November 2002 (Yang et al, 2020). SARS spread globally when an infectious physician from Guangdong, China, travelled to Hong Kong and transmitted the virus to family members, healthcare workers, and residents, some of whom travelled to other international destinations, leading to outbreaks in Singapore, Vietnam, and Canada. The SARS pandemic reached 32 countries– 8,422 people were infected and 919 died (Yang et al, 2020).

In Canada alone, 438 suspected and probable cases of SARS were reported, 44 people died, and 25,000 residents were placed in quarantine (Mazzulli et al, 2004) – most of whom were in the Greater Toronto Area (Low, 2004). The first reported case was a 78-year-old woman who came back from Hong Kong on 23 February 2003 and infected her family members and physician, eventually leading to outbreaks in various parts of the city. On 26 March 2003, SARS was declared a provincial emergency by the government of Ontario, allowing public health officials to impose quarantine measures on affected individuals (Mazzulli et al, 2004). Hospitals in the Greater Toronto

Area suspended non-essential services, limited visitors, created isolation units, and required protective clothing for exposed staff (Low, 2004).

The immense impact of the outbreak on affected individuals and their families, suspected cases, and the hospital system caused great public concern about contracting SARS, especially for those in Ontario (Blendon et al, 2004). Although public health officials exerted significant effort not to link ethnicity and illness, reports of the first SARS case in Canada, involving someone who had travelled from Hong Kong, coupled with initially limited public understanding of the mechanisms of disease transmission, led to many people unnecessarily avoiding Chinese businesses (Singer et al, 2003) and people suspected of having recently visited Asia (Blendon et al, 2004). These actions led to Chinese restaurants and Chinatowns being deserted (Hung, 2004), resulting in loss of income and livelihood. Moreover, Asian minorities, particularly those who look East or South East Asian, experienced discrimination, alienation, and harassment in various spaces, including public transportation, workplaces, and schools (Leung, 2008). For instance, interviewees in a study by Leung (2008) shared the following sentiments: "A lot of people avoided Chinese people. I know there were a lot of experiences regarding the subway. If you sneeze or cough, you could empty the train!" (Leung, 2008, p 138) and "When she did enter the foyer … she responded, 'Go back to your own country and stop transporting diseases here'" (Leung, 2008, p 139).

Although knowledge on the origins of SARS could have already led to the exclusion, avoidance, and overt harassment of various Asian minorities, these attitudes were further fuelled by the media calling SARS the 'Chinese plague' or the 'Oriental plague'. Media analysis in a report by Leung and Jiang (2004) revealed that 30 to 54 per cent of pictures in SARS-related articles from major Canadian magazines, such as the *National Post*, *The Globe and Mail*, and the Canadian edition of *Time*, depicted Asian people and/or Asian people wearing masks (Leung and Guan, 2004). North American editorial cartoons also closely associated the disease with drawings depicting Asian culture. These include a *New York Post* cartoon of people of East Asian appearance wearing a mask and getting swabbed in a 'SARS Testing Centre'; a *Tribune Review* cartoon showing an opened and smoke-releasing noodle take-away box labelled 'SARS' with a caption 'Bad Chinese Take-Out'; and an *International Herald Tribune* cartoon depicting the Great Wall of China with a 'SARS Quarantine' banner plastered by a person wearing a 'WHO' white coat (Hung, 2004). Media articles repeatedly emphasised the origin of the disease, including the names of the first patients and their racial and cultural background. For example, an article entitled 'Peril from the East' in the Canadian edition of *Time* mentioned the name of the first patient in Canada and described her as a 'super-spreader' who could have infected more than 155 people (Leung and Guan, 2004). Even the titles of articles, such as

'The SARS epidemic: the path; from China's provinces, a crafty germ breaks out' in *The New York Times* (Rosenthal, 2003), associated the virus with stereotypes of the Chinese as being cunning and/or sneaky. Such stereotypes were built on alleged unhygienic dietary practices of a particular group, with deep roots in colonialism (Ali, 2008). For instance, Rosenthal's (2003) article begins as follows:

> An hour south of Guangzhou, the Dongyuan animal market presents endless opportunities for an emerging germ. In hundreds of cramped stalls that stink of blood and guts, wholesale food vendors tend to veritable zoos that will grace Guangdong Province's tables: snakes, chickens, cats, turtles, badgers, frogs. And, in summer, sometimes rats, too. They are all stacked in cages one on top of another – which in turn serve as seats, card tables and dining quarters for the poor migrants who work there. (Rosenthal, 2003, p 1)

Even if an article's aim was to discuss the origins of the pandemic and to hypothesise possible sources of the virus, presenting information in a derogatory and culturally insensitive manner can risk perpetuating stereotypes and further encourage prejudiced views that may marginalise certain groups, such as Chinese Canadians. Participants in a qualitative study by Leung and Guan (2004) highlighted how the mainstream media sensationalised and racialised SARS. Some respondents stated: "The media kept talking about a specific travel pattern between Hong Kong, China and Toronto. I don't know if it was deliberate or necessary to keep saying that. This constant mapping of the disease, of course, kept attention on Chinese in Toronto as if everyone was suspect" (Leung and Guan, 2004, p 13), and "I feel that the SARS coverage overall was slightly biased and sometimes made Asian people look pretty disease-ridden. This was heightened in a culture in which racism, particularly against Asian communities, already permeates the sub-consciousness. It truly reinforced the feeling of 'other,' as an Asian person" (Leung and Guan, 2004, p 13).

Culturally sensitive communication (Airhihenbuwa et al, 2020) is crucial during pandemics, especially when lives and livelihoods are at stake. As important as it is to convey scientific facts, the way they are communicated, including the title of the article, the order and framing of the information presented, and the extent to which a particular negative attribute is linked to a racial/ethnic group, is equally important. As demonstrated by media portrayals and public reactions to them during the SARS pandemic, repeated and insensitive association of East Asian countries and people with the disease fuelled racially motivated discrimination and even harassment, causing people who appear East or South East Asian additional distress in an already distressing period.

COVID-19 control in Australia

The first cases of COVID-19, or Coronavirus Disease-19, were reported in Wuhan, China, on 31 December 2019. From the initial patients, the number of cases grew exponentially; and by 30 January 2020, the virus had spread to at least 19 countries outside China (Rothan and Byrareddy, 2020). The first positive case in Australia was confirmed on 25 January 2020 in a person returning from Wuhan (Andrikopoulos and Johnson, 2020), and the first cases of community transmission were reported on 2 March 2020 (Lupton, 2020). By November 2021, Australia had experienced three major waves of COVID-19, with the first wave associated with the initial worldwide spread of the virus, the second one from a breach in hotel quarantine protocols for people returning to the country (Coate, 2020), and the third through the incursion of the more contagious Delta variant (Jose, 2021). Lockdowns were implemented in all three waves and were successful in the first two, effectively reducing community transmission in Australia to zero. Lockdowns had limited effect in the third wave; however, new daily cases eventually started to drop as a result of increased population vaccination.

Similar to what was reported during SARS in Canada, Asian minorities in Australia experienced discrimination and harassment (Luo, 2020), considering that the first reported cases were in China and with the labelling of the disease as 'kung flu' and 'Chinese virus' by some prominent international politicians (Rogers, Jakes, and Swanson, 2020.Thus, communication on the pandemic has marginalised racial minorities once again. However, other forms of science communication, or lack of it, also caused distress and exclusion of minority ethnic groups, particularly those subjected to inequitable lockdown measures to manage outbreaks. Residential building lockdowns in Melbourne during the second wave and stricter restrictions for particular local government areas in Sydney during the third major wave illustrated how communication to racial minorities during pandemics needs to be done with them, ensuring that they also understand the scientific evidence behind pandemic control measures and giving them the opportunity to deliberate on the validity, appropriateness, and implementation of these protocols.

The second wave of COVID-19 was caused by a breach in hotel quarantine protocols for residents returning to the Australian state of Victoria from overseas (Coate, 2020). To stop the outbreak from growing, the Victorian government announced a hard lockdown for ten postcodes in the afternoon of 30 June 2020. The lockdown took effect at 11:59 pm on 1 July, preventing people from going outside their homes except for buying food and other essentials, care or caregiving, exercise, work, and/or study (ABC News, 2020a). On 4 July, 23 cases were detected in residential tower blocks, which led the Victorian Premier to implement an immediate full lockdown in nine public housing towers at 16:08 pm, preventing residents from leaving the towers for any reason. The immediate harsh lockdown was rationalised using scientific and medical evidence on the nature of viral transmission, particularly high transmission risk in overcrowded and communal facilities and the background health conditions of residents, which may put them at risk for worse outcomes if they get infected (ABC News, 2020b). To implement the lockdown, at least 500 police officers were deployed per shift. Most towers were released from the full lockdown on 9 July, except for the 33 Alfred Street tower, as 11 per cent of its residents were COVID-19 positive. Residents of the tower were required to isolate for nine more days (Fowler and Sakkal, 2020).

It is worth noting that a considerable proportion of the 33 Alfred Street tower residents prefer to speak a language other than English, including Somali, Arabic, Spanish, Amharic, Cantonese, Vietnamese, Oromo, Tigrinya, and Harari (Victorian Ombudsman, 2020). As such, it was important to use linguistically and culturally sensitive communication about the lockdown rules and their justification, especially with the lockdown's immediate implementation. However, an investigation by the Victorian Ombudsman on the tower lockdowns revealed that the Victorian Department of Health and Human Services acted in a wrong manner by:

> (a) failing to provide people at 33 Alfred Street including non-English speaking residents, with timely and accessible notice of the reasons for and terms of their detention (b) failing to notify people at 33 Alfred Street of the ability to complain about aspects of their treatment under section 185(1) of the Public Health and Wellbeing Act 2008 (Vic). (Victorian Ombudsman, 2020, p 179)

Moreover, it was only on the fifth or sixth day of the lockdown that some residents received translated materials regarding the purpose and the conditions of the lockdown. Given how abruptly the tower lockdown was implemented, there was not enough time to position qualified interpreters during the lockdown's critical first evening, potentially leading to confusion and distress for residents who were not proficient in English. Even the Victorian Chief Health Officer underscored that imposing the lockdown a day after it was announced, akin to the timing of the lockdown in the ten postcodes, could still have worked in managing the outbreak. It could have also provided the department enough time to consult with multicultural communities and for lockdown directives to be translated into multiple languages (Victorian Ombudsman, 2020).

The third major wave of COVID-19 in Australia started on 16 June 2021 when an airport limousine driver from Sydney's eastern suburbs tested positive for COVID-19. Daily new cases then started to increase, and by 26 June, 29 new daily cases were recorded. Stay-at-home orders were imposed on everyone in the Greater Sydney area (Taouk, 2021), allowing people to only leave home for essential shopping, medical care, exercise, or essential work/education (ABC News, 2021). Although most of the outbreaks were initially recorded in Sydney's east, new cases started to emerge in Sydney's west and south-west regions. Stricter restrictions were imposed on three local government areas (LGAs) in this region on 8 July (Smith, 2021), which were then extended to 12 LGAs on 12 August (Snow, 2021). Tighter measures in these LGAs included a curfew from 21:00 pm to 05:00 am, only one hour of outdoor exercise allowed per day, and fines for going outside an LGA of concern without a reasonable excuse (McPhee, 2021). Additional police and even Australian Defence Force personnel were deployed in these LGAs, with helicopters regularly hovering over houses and ordering people to stay home (Rachwani, 2021a). It is worth noting that the 12 LGAs of concern have significant populations of racial minorities and migrants, many of whom are essential workers who are not able to work from home. For instance, in the LGA of Canterbury-Bankstown, only 31.1 per cent of residents were born in Australia, and only 34.1 per cent speak English exclusively (Rachwani, 2021b).

Several issues with the state's response to COVID-19 in the 12 LGAs were raised by residents and concerned researchers, particularly the harsher restrictions imposed and the way they were communicated. For instance, in an interview with *The Guardian*, Dr Michael Camit of the South Western Sydney Local Health District highlighted how the government employed a paternalistic and top-down approach that did not account for community voices and the need for nuance (Rachwani, 2021b). This was echoed by criticisms from a resident of the government's focus on simply translating health information and directives, cherry-picking of community leaders to promote testing and vaccination, and lack of proactive building of relationships and trust with different communities (Rachwani, 2021b). These resulted in restrictions that did not account for the actual situation of people and households in these LGAs, such as the size of their home, nature of their work, and availability of local clinics for vaccination (Visontay and Taylor, 2021). Lack of engagement could also have led to the over-policing in the LGAs of concern, which made several residents feel intimidated and uncomfortable. Furthermore, there was a lack of transparency in police guidelines for checking the quarantine compliance of COVID-19 positive individuals and for monitoring the mental health impacts of their operations (Rachwani, 2021a). Finally, there was inconsistent use of scientific evidence and health advice for rationalising restrictions. The implementation of a curfew contrasted with initial statements from the Premier and Deputy Premier on the ineffectiveness of curfews, limited evidence to support them, and their impact on shift workers. The LGAs declared as hotspots were also contested, as LGAs with growing case numbers and lower vaccination rates were not classified as LGAs of concern, while LGAs with sharply declining active cases were still subjected to the stricter restrictions (Baker and Wade, 2021). The New South Wales government also ignored health advice from its chief health officer on applying the same lockdown measures across greater metropolitan Sydney, a measure that could have had a better chance of driving down case numbers and could have avoided stigmatising communities in Western Sydney (Nilsson, 2021).

Overall, mistakes in the way science was communicated to and with racial minorities, migrants, and refugees in the towers in Melbourne that were locked down were repeated with people in LGAs in Sydney that had stricter restrictions. It seemed like lessons were not learned, and that a stronger voice is needed to underscore the importance of an inclusive and participatory approach in pandemic science communication. Both examples demonstrate the value of timely information translation, acknowledging socio-economic and cultural contexts (Airhihenbuwa et al, 2020) and accounting for the effects of structural racism (Rachwani, 2021b). Both cases also demonstrate the need for greater transparency on the science, both natural and social, that was used in developing pandemic policies. Government leaders must be open to questioning and interrogating the science used in formulating policies, and they must be willing to revise regulations in accordance with epidemiological data, ethical and social justice considerations, and insights from inclusive public deliberation.

Conclusion: a participatory pathway forward

Communication regarding the SARS pandemic's origins and how it reached Toronto closely associated Asian minorities with the disease, leading to stigmatisation, discrimination, and even harassment. On the other hand, lack of timely and proper communication with racial/ethnic minorities in the design and implementation of lockdowns to manage COVID-19 in residential and local government areas in Sydney and Melbourne negatively affected their well-being. These scenarios demonstrate why science communication on and to minority ethnic groups needs to be done with them.

Airhihenbuwa et al (2020) highlight the importance of culture in communication and engagement initiatives during COVID-19 and how efforts should strive to understand local contexts, seeing beyond the 'deficits' and acknowledging the persistence and resilience of racial minority communities. Hyland-Wood et al (2021) also recommend the consideration of diverse communities in communications and interventions during pandemics. Drawing from and building upon these suggestions, journalists and science communicators should closely engage with racial minorities when communicating about pandemics, given the potential of what is communicated to adversely impact the safety, privacy, well-being, sense of belonging, and livelihood of particular ethnic groups. Moreover, policy makers should move away from a top-down or paternalistic approach to one that closely engages (Viaña et al, 2021), solves problems, and brokers with communities (Orthia et al, 2021) to address issues that ultimately affect them, including but not limited to viral outbreaks. Science communicators should always acknowledge the value of the perspectives and lived experiences of racial and ethnic minorities, including their ability to speak to both their and the broader public's interests (Viaña et al, 2021). This will ensure that pandemic communication and response are conducted in an equitable manner, whereby minorities are recognised, respected, and included. Only by empowering racial minorities and other underrepresented social groups can we truly address the health disparities they experience and properly cultivate a science communication landscape that is not just informative but also justice-oriented.

References

ABC News (2020a) 'Victorian coronavirus stay-at-home orders reimposed across Melbourne hotspot suburbs', ABC News, 30 January, Available from: https://www.abc.net.au/news/2020-06-30/victoria-coronavirus-hotspot-local-lockdowns-in-melbourne/12407138

ABC News (2020b) 'Victoria coronavirus cases rise by 108 as Daniel Andrews strengthens lockdown at nine public housing estates', ABC News, 4 July, Available from: https://www.abc.net.au/news/2020-07-04/victoria-coronavirus-cases-rise-by-108-lockdown-new-postcodes/12422456

ABC News (2021) 'Greater Sydney's two-week lockdown – this is what you need to know about the new restrictions in the capital and around NSW', ABC News, 27 June, Available from: https://www.abc.net.au/news/2021-06-26/nsw-covid-19-lockdown-rules-explained/100246644

Airhihenbuwa, C.O., Iwelunmor, J., Munodawafa, D., Ford, C.L., Oni, T., Agyemang, C. et al (2020) 'Culture matters in communicating the global response to COVID-19', *Preventing Chronic Disease*, 17: E60.

Ali, S.H. (2008) 'Stigmatized ethnicity, public health, and globalization', *Canadian Ethnic Studies*, 40(3): 43–64.

Andrikopoulos, S., and Johnson, G. (2020) 'The Australian response to the COVID-19 pandemic and diabetes – lessons learned', *Diabetes Research and Clinical Practice*, 165, 108246, Available from: https://doi.org/10.1016/j.diabres.2020.108246

Asian Australian Alliance and Chiu, O. (2020) 'Covid-19 coronavirus racism incident report: reporting racism against Asians in Australia arising due to the COVID-19 coronavirus pandemic', Diversity Arts Australia, Available from: http://diversityarts.org.au/app/uploads/COVID19-racism-incident-report-Preliminary-Official.pdf

Baker, J. and Wade, M. (2021) 'Suburban fury: Delta divide deepens over Sydney hotspot lockdowns', *The Sydney Morning Herald*, 18 September, Available from: https://www.smh.com.au/national/nsw/suburban-fury-delta-divide-deepens-over-sydney-hotspot-lockdowns-20210916-p58s6i.html

Blendon, R.J., Benson, J.M., DesRoches, C.M., Raleigh, E., and Taylor-Clark, K. (2004) 'The public's response to Severe Acute Respiratory Syndrome in Toronto and the United States', *Clinical Infectious Diseases*, 38(7): 925–31.

Coate, J. (2020) 'COVID-19 hotel quarantine inquiry interim report and recommendations', Victorian Government Printer, 6 November, Available from: https://www.quarantineinquiry.vic.gov.au/sites/default/files/2020-11/COVID-19%20Hotel%20Quarantine%20Inquiry%20Interim%20Report%20and%20Recommendations%206%20November%202020.pdf

Fowler, M. and Sakkal, P. (2020) 'One public housing tower to remain in lockdown, restrictions to ease on others', *The Age*, 9 July, Available from: https://www.theage.com.au/national/victoria/one-public-housing-tower-to-remain-in-lockdown-as-restrictions-on-others-to-be-eased-20200709-p55amh.html

Hung, H. (2004) 'The politics of SARS: containing the perils of globalization by more globalization', *Asian Perspective*, 28(1): 19–44.

Hyland-Wood, B., Gardner, J., Leask, J., and Ecker, U.K.H. (2021) 'Toward effective government communication strategies in the era of COVID-19', *Humanities and Social Sciences Communications*, 8(1): 30.

Jose, R. (2021) 'Australia's daily COVID-19 cases near 2,000 as Delta gains ground', Reuters, 10 September, Available from: https://www.reuters.com/world/asia-pacific/covid-19-cases-rise-australias-victoria-regions-exit-lockdown-2021-09-10/

Keil, R. and Ali, H. (2006) 'Multiculturalism, racism and infectious disease in the global city: the experience of the 2003 SARS outbreak in Toronto', *TOPIA: Canadian Journal of Cultural Studies*, 16: 23–49.

Leung, C. (2008) 'The yellow peril revisited: the impact of SARS on Chinese and Southeast Asian communities', *Resources for Feminist Research*, 33(1/2): 135–49, 155.

Leung, C. and Guan, D. (2004) 'Yellow peril revisited: impact of SARS on the Chinese and Southeast Asian Canadian communities', Toronto: Chinese-Canadian National Council, June, Available from: https://www.academia.edu/919335/Yellow_peril_revisited_Impact_of_SARS_on_the_Chinese_and_Southeast_Asian_Canadian_communities

Low, D.E. (2004) 'SARS: lessons from Toronto', in S. Knobler, A. Mahmoud, S. Lemon, A. Mack, L. Sivitz, and K. Oberholtzer (eds) *Learning from SARS: Preparing for the Next Disease Outbreak; Workshop Summary*, Washington, DC: National Academies Press, pp 63-71, Available from: https://www.ncbi.nlm.nih.gov/books/NBK92467/.

Luo, L. (2020) 'Survey: anti-Asian racism a problem on Australian streets', UOWTV Multimedia – Social First Newsroom, 20 July, Available from: https://www.uowtv.com/2020/07/20/racial-abuse-against-asian-australians-during-the-covid-19-pandemic/

Lupton, D. (2020) 'Timeline of COVID-19 in Australia: the first year', Medium, 13 August, Available from: https://deborahalupton.medium.com/timeline-of-covid-19-in-australia-1f7df6ca5f23

Mazzulli, T., Kain, K., and Butany, J. (2004) 'Severe Acute Respiratory Syndrome: overview with an emphasis on the Toronto experience', *Archives of Pathology & Laboratory Medicine*, 128(12): 1346–50.

McPhee, S. (2021) 'Mandatory masks and curfew for LGAs of concern: new COVID-19 restrictions for NSW', *The Sydney Morning Herald*, 20 August, Available from: https://www.smh.com.au/national/nsw/sydney-lockdown-extended-masks-mandatory-curfews-in-areas-of-concern-20210820-p58kg1.html

Nilsson, A. (2021) 'NSW government "ignored health advice" from Kerry Chant, new emails show', News.com.au, 22 November, Available from: https://www.news.com.au/national/nsw-act/news/nsw-government-ignored-health-advice-from-kerry-chant-new-emails-show/news-story/b936a21c241de654e0b7e347486a1581

Orthia, L.A., McKinnon, M., Viana, J.N., and Walker, G. (2021) 'Reorienting science communication towards communities', *JCOM: Journal of Science Communication*, 20(3): A12.

Pan, D., Sze, S., Minhas, J.S., Bangash, M.N., Pareek, N., Divall, P. et al (2020) 'The impact of ethnicity on clinical outcomes in COVID-19: a systematic review', *EClinicalMedicine*, 3(23): 100404.

Patrick, A. (2021) '"We tried everything" to avoid lockdown', Financial Review, 8 July, Available from: https://www.afr.com/policy/health-and-education/we-tried-everything-to-avoid-lockdown-20210708-p5881w

Rachwani, M. (2021a) '"We feel intimidated": residents in south-west Sydney Covid hotspots say police are making things worse', *The Guardian*, 15 August, Available from: https://www.theguardian.com/australia-news/2021/aug/15/we-feel-intimidated-residents-in-south-west-sydney-covid-hotspots-say-police-are-making-things-worse

Rachwani, M. (2021b) 'Fear and loathing in western Sydney: how NSW's Covid response failed migrant communities', *The Guardian*, 13 August, Available from: https://www.theguardian.com/australia-news/2021/aug/13/fear-and-loathing-in-western-sydney-how-nsws-covid-response-failed-migrant-communities

Reincke, C.M., Bredenoord, A.L., and van Mil, M.H. (2020) 'From deficit to dialogue in science communication', *EMBO Reports*, 21(9): e51278.

Rogers, K., Jakes, L., and Swanson, S. (2020) 'Trump defends using "Chinese virus" label, ignoring growing criticism', *New York Times*, 18 March, Available from: https://www.nytimes.com/2020/03/18/us/politics/china-virus.html

Rosenthal, E. (2003) 'The SARS epidemic: the path; from China's provinces, a crafty germ breaks out', *New York Times*, 27 April, Available from: https://www.nytimes.com/2003/04/27/world/the-sars-epidemic-the-path-from-china-s-provinces-a-crafty-germ-breaks-out.html

Rothan, H.A. and Byrareddy, S.N. (2020). 'The epidemiology and pathogenesis of coronavirus disease (COVID-19) outbreak', *Journal of Autoimmunity*, 109, 102433, Available from: https://doi.org/10.1016/j.jaut.2020.102433

Singer, P.A., Benatar, S.R., Bernstein, M., Daar, A.S., Dickens, B.M., MacRae, S.K. et al (2003) 'Ethics and SARS: lessons from Toronto', *BMJ*, 327(7427): 1342–4.

Smith, R. (2021) 'Southwest Sydney emerges as NSW's new Covid-19 ground zero', News.com.au, 8 July, Available from: https://www.news.com.au/national/nsw-act/news/southwest-sydney-emerges-as-nsws-new-covid19-ground-zero/news-story/fccc75d4647a0275e147611c2bf3adb2

Snow, D. (2021) 'Delta's dawn: our winter of discontent', *The Sydney Morning Herald*, 28 August, Available from: https://www.smh.com.au/national/nsw/delta-s-dawn-our-winter-of-discontent-20210827-p58mjc.html

Tan, M.K.I. (2021) 'COVID-19 in an inequitable world: the last, the lost and the least', *International Health* [online ahead of print] ihab057. doi: 10.1093/inthealth/ihab057

Taouk, M. (2021) 'From Bondi to Wollongong: tracking Sydney's latest COVID-19 outbreak', ABC News, 20 August, Available from: https://www.abc.net.au/news/2021-06-23/nsw-covid-bondi-outbreak-timeline/100237834

Viaña, J.N., Raman, S., and Barber, M. (2021) 'From paternalism to engagement: bioethics needs a paradigm shift to address racial injustice during COVID-19', *The American Journal of Bioethics*, 21(2): 96–8.

Victorian Ombudsman (2020) 'Investigation into the detention and treatment of public housing residents arising from a COVID-19 "hard lockdown" in July 2020, 2020', Victorian Ombudsman, December, Available from: https://assets.ombudsman.vic.gov.au/assets/Reports/Parliamentary-Reports/Public-housing-tower-lockdown/Victorian-Ombudsman-report-Investigation-into-the-detention-and-treatment-of-public-housing-residents-arising-from-a-COVID-19-hard-lockdown-in-July-2020.pdf?mtime=20201216075340

Visontay, E. and Taylor, J. (2021) 'Tougher Covid restrictions for western Sydney criticised for threatening wellbeing of state's poorest', *The Guardian*, 21 August, Available from: https://www.theguardian.com/australia-news/2021/aug/20/tougher-covid-restrictions-for-western-sydney-criticised-for-threatening-wellbeing-of-states-poorest

Yang, Y., Peng, F., Wang, R., Guan, K., Jiang, T., Xu, G. et al (2020) 'The deadly coronaviruses: the 2003 SARS pandemic and the 2020 novel coronavirus epidemic in China', *Journal of Autoimmunity*, 109: 102434.

3

Breaking the Silos: Science Communication for Everyone

Amparo Leyman Pino

Introduction

It has been over a decade since science centres raised concerns about the need to address the changes in the population demographics of the United States. These changes are also a growing reality in other regions of the Global North, such that in the United Kingdom (UK), the 2021 Census data show that a number of major Cities are now 'minority-majority' populations – meaning there is no longer a White majority (UK-ONS, 2023; Vinter, 2023). So-called ethnic 'minorities' in societies that are mainly White now represent a significant proportion of the population, accounting for nearly 20 per cent of the population of the United States, for example, and in many US states the percentage is even greater. In the State of California, for example, the Latin American diaspora represents nearly 40 per cent of the population (US Census Bureau, 2019).

Science centres and museums wish to remain relevant in their societies and to be considered the go-to places for informal science learning. With the growing shift in racial demographics in the Global North, these institutions are increasingly questioning their relevance among 'racialised minorities', 'underrepresented communities', and 'communities at-promise' as they seek to be more inclusive, aiming to engage *all* citizens in a more effective way.

Often their solution is to create and customise programmes and exhibits that cater specifically to these groups. Even though this may well seem to be a great idea, in the long run it has created silos instead of engendering a sense of community, a celebration of societal diversity, and an opportunity for diverse groups to learn from one another. In addition, a critical goal of inclusion in science centres and museums should be that of providing learning environments where all people, regardless of their situation or attributes, are welcomed and able to play, learn, and engage. As humans, we possess a wide diversity of attributes and situations: age, culture, gender identity, nationality, mental and physical abilities, language, religion, education, work experience, political views, income, and so on.

It is time to adopt an inclusive approach in the design of informal science learning experiences, inside and beyond the walls of these science centres and museums. We need to challenge ourselves and the field and think of ways to offer a science communication agenda for all. This chapter will present good practice strategies and transformational case studies on how to break these silos and thus, foster inclusive science learning and communication experiences across the race, gender, language, and socio-economic divides in societies in the Global North.

Frames for inclusive science communication: science communication and the democratisation of science

Science communication is the practice of informing, educating, raising awareness of science-related topics, and increasing the sense of wonder about scientific discoveries and arguments. Dawson (2013) explains that beyond building a bridge between the people involved in scientific research and different groups, science communication is 'involved in developing government science policies, understanding relationships between the "public" and "scientists", and creating science stories in the mass media, as well as exploring how people learn about and engage with science'.

In Spanish, the term is called *popularización de la ciencia*, meaning making science popular, from the Latin *popularis*, related to the people, hence giving ownership of science to people, making science accessible to the public, to all the people, democratising science. Consequently, science is not an activity limited to those who do science or scientists by career or professional background. When referring to the 'popularisation of science', in an ideal world the ownership of science is related to the whole process: from scientific research, learning how to communicate such research, writing books and even comics, bringing this knowledge to science centres and transforming it into a hands-on experience, to television and radio, and placing it at the centre of a conversation that explains to us how we relate to it in our everyday life. In his 2013 lecture at the Centre for Science and Policy, Sir Mark Walport, a former British Government Chief Scientific Advisor, said that "science isn't finished until it's communicated"; science communication is thus part of the whole process of the scientific endeavour.

In its ultimate exposition, scientific knowledge doesn't stay in laboratories, universities, and/or complex books or peer-reviewed articles but is something shared with all of us. The inherent challenge for science communication is to make it accessible to us all, regardless of the language we speak, our cultural background, the habits and customs we practise, the religion we believe in, the places where we are located, our gender, our race, and so on.

Breaking the silos: science for everyone

This chapter is focused on providing new perspectives for science communication in the breaking of barriers and silos, where we can draw together different strategies that elaborate diverse approaches and frameworks for engaging those groups that have been excluded. These scenarios need to encapsulate an explicit intention for inclusion, clarity on intentions, goals, and methods, and frame a welcoming culture to provide science communication experiences for all.

There is a critical need to move the inclusive science communication agenda to a place of action and to propose strategies to address the 'how', specifically how we are going to break the silos when we are trying to include audiences that have been excluded, inculcate narratives that are not part of the Western understanding of science, and engage communities that do not feel welcomed or able to relate to the places, spaces, and content of science centres and museums. These challenges call to mind numerous situations when institutions initiate conversations with and even bring on board professionals as consultants/advisors who are part of the communities they want to include and then request their expertise in designing programmes for a specific target audience or creating a science communication programme related to cultural traditions from those communities. The net result of these well-intentioned efforts is that they end up creating silos within their normal programmatic offering instead of thinking of how to adapt and integrate the knowledge, language, culture, and preferences of these excluded groups so as to include them in the overall content and delivery of the institution's science communication activities, programmes, and engagement.

The recommendation here is that science centres and museums should desist from narrow, silo-mentality approaches to inclusion through undertaking initiatives, which though well intentioned, end up serving specific audiences, thereby provoking segregation and disconnection between people and disconnections from their regular offerings and the operation of their institution's mission. This is a call for these institutions to be intentional and think strategically about how they can design, prototype, and offer truly inclusive science communication through programmes, exhibits, festivals, television shows, online learning, and so on, to mention some examples of the various means for the popularisation of science. This is an invitation to create space for everyone, in an intentional way, and to have clarity internally and externally as to how institutional approaches will work to dismantle the barriers and thus create welcoming spaces for all.

There is an unfortunate tendency for institutions to take short-cuts in order to achieve 'quick wins', meaning undertaking practices for the integration of audiences that have very short-lived and limited impact instead of aiming for an authentic inclusion of diverse audiences and communities, which has a longer-lasting and more sustainable impact. Thus, it is common for science centres and museums to create programming intended for particular community groups, but when relied upon as the solution to inclusion, institutions fail at diversifying, community-building, and transforming. To work on a larger scale and achieve transformation, these institutions need to focus their institutional capacities and resources on integrating knowledge holders from underserved communities, including their language, culture, and preferences, so that the science centre or museum and its contents can truly be seen to belong to the diverse communities it purports to serve.

This chapter also seeks to highlight good practices and successful transformation strategies. A number of science centres and museums have been successful in breaking down barriers to 'whole-scale' inclusion, and when that happens, space is created for members of the public who have been excluded or siloed to engage with science centre staff, exhibits, content, and other members of the public. When barriers end, new scenarios begin – scenarios that look, feel, and sound like explicit intentions for inclusion with clear goals and methods, where the science centre culture is a shared culture that welcomes and provides science communication experiences for everyone.

Transformative inclusion in the science communication field will only come from collective action and leadership and collective reflection – with institutional, individual, and collective insights into blind spots, biases, systemic exclusion, and a radical commitment to the unlearning of poor historical and contemporary practices. Included in this dimension is the collective strategy linked with respectful enthusiasm to learn from and to include underserved minority community knowledge holders that have been excluded from narratives of Western science and science centres.

From integration tactics to inclusion strategies

The act of diversifying audiences requires an internal effort to adapt the current offering and make it available, relatable, participatory, and connected to as many people as possible, especially to the new audiences the institution wants to reach.

Some US institutions attempt to solve this challenge by creating programmes, events, or exhibits that are targeted to a specific group: for example, it is common to find institutions hosting targeted events during special cultural highlights of minority communities, such as the *día de muertos* (the day of the dead), to attract people from the Mexican diaspora. The duration of these events usually ranges from one to seven days. In most cases, for the rest of the year, this community is neglected, and the institutions do not bother to invest additional time or undertake follow-up or extension activities to further expand, deepen, and up-scale their connections with the community and nurture the relationships built during the preparation and hosting phases of these activities. Even though they are a good way of initiating community engagement, these types of engagements often create segregation and are doomed to fail due to their very limited and patchy impact. People who don't identify with the cultural event either don't attend it or, if they do attend, don't feel included because it is delivered in another language or hasn't been contextualised or explained in a manner that is multicultural in scope. On the other hand, the underserved minority community that has been invited to the institution to celebrate their culture end up excluded as well, as the 'special' event becomes a silo or bubble that limits their participation in the institution's regular programmatic offerings. Another example that comes to mind is the 'free community days' aimed at low-income families, which enables them to attend but who often find themselves 'segregated' in the venue, because the regular fee-paying patrons do not want to interact with them.

Other examples of the limiting nature of integration tactics include initiatives to engage new audiences into the institution's constituency through having some exhibits that are bilingual or creating programmes for girls in STEM (science, technology, engineering, mathematics). These ephemeral programming scenarios may bring these audiences closer, but they will remain corralled in their own silo without fully participating with the mainstream audiences or enjoying the same experiences as the institution's other visitors. Integration is a good first step towards inclusion, better than one-off stand-alone events that occur once a year with a marginalised group, but it is still far from being an inclusive practice, and it ultimately fails to build a diverse community as it does not enable marginalised groups to be fully incorporated into the framework of systematic organisation-wide practices.

Inclusion in science centres and museums is the art of intentionally designing learning, communication, and interactive experiences whereby people from every given background, ability, culture, language, socio-economic status, and level of education can participate on a par with the mainstream audience. One premise of inclusive learning environments is the acknowledgement that every individual learns in a particular way and that the system needs to adapt to each learner, with 'the commitment to removing all barriers to the full participation of everyone as equally valued and unique individuals' (Eid, 2016). The informal science education field has become proficient in translating complex scientific concepts into hands-on experiences, meaning that it is expert at making science accessible to the public. This expertise needs to transcend further into inclusive approaches that enable science communicators to think of innovative ways to offer tangible and experiential learning that everyone can participate in, enjoy, and understand.

Inclusive science communication: good practice case studies

The strategic goal of engaging diverse audiences has been an objective that many informal learning institutions have been trying to adopt for some time. Conversations on how to provide learning experiences that can be enjoyed by all kinds of visitors are still ongoing. This is a never-ending endeavour that needs more practitioners and professionals in the field to further enrol with this movement and contribute their creative thinking to the conceptualisation of good practice strategies that will shift both institutional and field-wide cultures and practices and enhance inclusive learning environments.

The following good practice case studies, taken from transformative developments in some science centres and museums in Europe and the United States, should hopefully serve as signposts, inspiration, and encouragement to practitioners and professionals in the field, regardless of their timeline on the journey to transformation in their institutions. It is also critical in this regard for practitioners, no matter where they are on the spectrum of experience, to understand from these case studies that they are not alone in their quest for transformative inclusion and that there are other educators and science communicators like them who are reflecting, designing, prototyping, and putting into practice great ideas that they can learn from and potentially adapt to their particular national contexts, circumstances, and demographic scenarios.

In the interests of transparency and institutional accountability, only the science centres and museums that have produced publicly available documentation on their diversity and inclusion work as highlighted in this chapter have been named.

Mission-driven intentionality

Inclusiveness is a quality, a value, a practice, a style, a way of doing things, and an institutional goal. The institutions that are authentically inclusive are those that practice inclusivity as their core functionality: it is clearly stated in their mission, their vision, articulated in their core values, and visible in their practices.

For example, one European science centre has adopted a unique approach in that it has made the transition towards inclusive practices a strategic institutional policy framework that must be complied with by all staff members, regardless of their role in the centre. This mandatory institutional requirement thus means that inclusion is not an exclusive responsibility of a specific department, such as the education and/or exhibits departments, but is an institution-wide policy mandate that impacts the way it designs inclusive learning experiences, communicates these through empowering marketing campaigns, and articulates it in its rationale for funding. For this European science centre, accessibility is a must, and it is placed at the heart of all its projects, from the very beginning. Accessibility is considered and thought through in dialogue with all departments and all relevant stakeholders. The dialogue interactions aim to ensure that the centre's projects do not discriminate against or exclude anyone and that visitors have equitable access to all the different services and offerings without any restrictions, safely and regardless of their abilities.

Another good practice exemplar is that of explicitly formalising the institutional ideals, vision, mission, and drive for greater inclusion by putting these in writing. Once this is stated, the science centre is then driven to create capacity in its staff in order to deliver on the imperative of making this publicly positioned document a functioning reality. Some science centres have found this practice to be a fundamental step in transforming an ideal into a reality and a policy framework into a practice framework. More importantly, the dialogue and consensus-building processes leading up to the conceptualisation, framing, and adoption of these transformative statements and policies can be equally inclusive and transformative of institutional cultures, making these processes very much worthwhile in themselves. The science centre board, staff, volunteers, patrons, community partners, and other stakeholders that have participated in these processes to create and approve such policies are also then able to act as key oversight monitors, to hold the institution to account, and ensure the sustainable delivery of these good practices once they have been put in place. For example, the Monterey Bay Aquarium (2019) in the United States issues an annual report where it analyses the demographic shifts and its achievements and improvements regarding the diversification of its audiences. The report also highlights the strategies it has undertaken to strengthen its relationships with local communities through research projects, programmes, and exhibits. This annual report is a critical tool and platform that the Aquarium uses to make itself accountable to its mission and commitment to becoming an inclusive institution.

Inclusive learning environments

Inclusive learning environments intentionally create space for everyone through the processes of facilitating clear institutional recognition both internally and externally that some groups are unable to participate. Through these processes, science centres and museums are able to ensure that they are addressing the barriers – dismantling them – to create spaces for a diverse range of visitors and audiences. However, a remaining challenge for many science centres and museums is how they can then aim to develop exhibits and programmes that can inclusively communicate the same message and provide the same learning experiences for everyone, given the wide disparities in human attributes, such as gender, race, and language.

A European science centre tackled this challenge with a mind-set that focused on what people are able to do and identifying what they cannot do – the barriers – so that they could then think of ways of diminishing or overcoming these barriers. The centre assessed its activities and interactive experiences as being potentially difficult or challenging for a wide range of human attributes, such as people who do not speak the official language, people with visual or auditory impairments or physical/mental disabilities, and then creatively worked on normalising the interactive experience so that it could be the same for everyone through removing identified and potential barriers and also offering other alternatives. This transformative approach has led this European science centre to adopt a unique common practice when designing inclusive learning experiences – that of making the interaction and thus, the experience, multisensory, meaning it incorporates all the senses so that a balance is achieved for all visitors across multiple attributes. For example, the centre makes use of tri-dimensional models, where visitors who are visually impaired can have a tactile reference and so can the rest of their audiences too. The interactive learning experience from each exhibit is structured to be coherent and similar for all visitors, in an inclusive design paradigm.

Universal design principles

The 7 Principles of Universal Design (2020a) are a useful guideline to consider when designing and prototyping inclusive learning experiences. These seven principles are recommended as a checklist or for assessment purposes: for example, Principle 2, 'Flexibility of Use', suggests that flexibility be incorporated in the design phase in order to accommodate both right- or left-handed access and use. This simple consideration in design and planning is one that could have a positive impact on the diversity of audiences and visitors. The US National Disability Authority (2020b) explains it clearly when defining universal design:

> [T]he design and composition of an environment so that it can be accessed, understood and used to the greatest extent possible by all people regardless of their age, size, ability or disability. An environment … should be designed to meet the needs of all people who wish to use it.

The 7 Universal Design Principles are as follows:

1. Equitable Use;
2. Flexibility of Use;
3. Simple and Intuitive Use;
4. Perceptible Information;
5. Tolerance for Error;
6. Low Physical Effort; and
7. Size and Space for Approach.

These principles highlight the importance of avoiding the segregation or stigmatisation of users and the need to be careful in avoiding the use of any condescending language, attitudes, or gestures when communicating with audiences who have struggled with discrimination because of their characteristics or attributes, hence the importance of staff capacity building, whether they are directly involved or not, in the creation of learning experiences and in interactions with the public. Cultural sensitivity and accessibility competences need to be considered in the on-boarding processes and in the continuous professional development of all staff and practitioners, regardless of where their specific roles in the science centre or museum are located.

Start from what you have

Every science centre or museum has some existing strengths, successful exhibits, programmes, and relationships with their local communities. These should provide a strong foundation and springboard from which to reach for higher levels when trying to become more inclusive and diverse. Assessing the current offerings of the organisation and thinking of creative ways to adapt current exhibits and programmes to make them more accessible is a great way to start and a solid step to broadening the audiences who can participate in the institution's learning spaces.

When the Monterey Bay Aquarium wanted to create programmes for the Latin American diaspora, it initially considered the idea of 'segregating' them by offering a programme tailored to them specifically. However, when this well-intentioned but limited idea was challenged, it came up with the suggestion of adapting its very successful live presentations to all audiences regardless of the language they spoke. The Aquarium called this 'blended language programming' (Leyman Pino et al, 2020). Blended language consists of one narrative (here, in two languages, Spanish and English) that switches from one language to the other, without echoing the messages, which makes it equally engaging to audiences who only speak Spanish, those who only speak English, and those who are bilingual, all at the same time and in the same space. This is an inclusive approach that requires human capital and creativity to design from a place of empathy and connection.

Another interesting tool for inclusion is the use of audio recordings that narrate the content of the exhibition for people who are visually impaired, pre-readers, or those who are auditory learners. Adaptation is a powerful method to scaffold from a solid foundation and iterate to find the sweet spot for inclusion.

The chain of accessibility

Creative thinking with regard to inclusion strategies should not relate only to exhibition halls or be limited to what science centres and museums can do in situ or in the immediate environs of their buildings. It is important to go further and take into account the holistic concept of the 'chain of accessibility'. The chain of accessibility concept refers to the incremental and multi-level steps visitors need to take from their place of living to the science centre or museum and back. This principle is critical to ensuring that these inclusive learning experiences are accessible to all the people from any community, meaning visitors do not have any barriers in commuting from their homes to the science centre, museum, or learning institution. This good practice approach entails a door-to-door analysis, in terms of inclusive access, for a range of underserved visitors, such as those from low-income neighbourhoods, people who have physical mobility challenges, or those who are visually impaired. All these groups will experience challenges in one form or another in getting from their homes to the science centre and back, and these needs will have to be considered and creatively managed by the institutions for inclusive access and participation to be engendered in truly holistic ways and means.

The chain of accessibility principle means that institutions should not only consider their ticket entry price as a factor for accessibility but also how people get to where their exhibits are and how long it takes them to do so. Some institutions establish initiatives such as 'free days' or 'pay as you can', which can be useful, but they often forget to bring on board additional financial considerations such as the cost of transportation to and from their venue and other logistical parameters such as whether their venue is easy to get to from the nearest bus or subway station. There are some places where public transportation is not efficient or not even accessible for people in wheelchairs, for example.

Delivering on the good practice imperatives of the chain of accessibility principles requires a holistic approach to addressing accessibility needs before visitors arrive at the science centre or museum, during their visit, and after they leave – in other words, the whole cycle of their experience, engagement, and interaction. However, this holistic framework requires the involvement of local governments and other public and private sector stakeholders working together to establish, facilitate, and sustain a seamless chain of accessibility for underserved groups to be fully enabled to inclusively access and engage with science centres and museums in their localities and communities.

Designing inclusive learning experiences and environments

The previous section showcased good practice examples of how to adapt current offerings. In this section, the good practice case studies will focus on the tools and methodologies needed when creating inclusive learning experiences and environments from scratch.

Integrating the key people and stakeholders

The creation of inclusive programmes and exhibits entails multidisciplinary and multi-stakeholder teams comprising graphic and industrial designers, educators, science communicators, and scientists and/or experts in the topic working together in seamless synergies. In addition, and in order to further assure the accessibility and inclusiveness of the outcomes, the in-house team in the science centre or museum will need to engage and involve external experts such as community-based organisations and diverse members of specific underserved groups based on criteria such as age, race, gender, and ability. All these strategies, methods, goals, and outcomes need to be communicated internally to all members of the institution as well as to all its constituents, partners, stakeholders, and potential visitors. Becoming an inclusive institution is a non-stop endeavour, and science communicators should always be looking for novel areas of improvement and innovative opportunities to do even better.

As an example, one American science centre, in planning a new exhibition, chose to invite several community partners that would bring a wide array of perspectives to the table. The exhibition was conceptualised to target Spanish-speaking girls in the 9–14 age group, and so the science centre partnered with an organisation that supports Spanish-speaking families with daughters in that age group. The science centre sought to empower these communities by asking them to help the institution with their ideas for the creation of an exhibition that would engage all girls, but from the unique perspectives of *their* identity, experience, and reality. These Spanish-speaking community organisations were key to the design process from their own sociocultural lens, and yet at the same time, they were also challenged to think what all girls might like and cherish in such an exhibition. On the other hand, the science centre also partnered with an organisation that serves families of refugees and immigrants so that they could learn different points of view from the multiple sociocultural perspectives and experiences of a more diverse group. A third strategy was the inclusion of scientific experts in the topic from other diverse race and gender backgrounds whose input ensured that the ideas from all the groups are relatable to all people, as well as being scientifically validated. With these multiple inclusive approaches, the science centre was then able to refine, coalesce, and frame the exhibition across the intersections of all these common meeting points, synergies, and consensus on interests through careful listening. The end result was that they were then able to produce a uniquely inclusive exhibition that connected with a large cross-section of their local community.

Inclusion rooted in community partnerships

Science centres and museums should be mindful that their purpose in co-creating with communities should be premised on building deep, empowering, respectful, and long-term relationships with them and not just as a one-time interaction for the sake of a programme or exhibit. This good practice framework entails a continuous two-way conversation that enables these institutions to understand how to include diverse and underserved voices and communities, identify their needs/preferences and the most effective channels of communication for engaging with them, and how to prototype ideas, troubleshoot with them, and run projects inclusively with them – in other words, develop genuinely transformative and sustainable partnerships with these communities. As a result, they will become radically inclusive institutions that are better able to serve all of their local communities.

Institutions often define this approach as 'community outreach', while others prefer to use the phrase 'community partnerships', because this places the institution and the community at the same level of inputs and responsibility, as equals. Community partnerships start where the communities mingle: the community centres, the five or six blocks on that part of the neighbourhood, at the temple, the mosque, the church, grocery shops, the Sunday football tournament, and so on – these are the places where science communicators need to show up, interact, engage, and gain the trust of their local communities as partners for inclusion and diversity. It is easier to enter these circles through the endorsement of a person or organisation that is valued and respected in the community; nonetheless, science communicators need to win the respect of the whole community too.

Another good practice to consider when engaging in partnerships with communities, especially in the context of 'picking their brains', is that of ensuring that considerations are given to some kind of financial compensation, remuneration, or honorarium. Institutions should ensure that budgets are set aside to potentially cover the salaries, stipends, or tokens of gratitude for the contributions of community members in recognition that their time, knowledge, expertise, and dedication should be valued, compensated, and remunerated in line with other sources and knowledge contributors. In the Legacy Document from the exhibit 'Roots of Wisdom', the Oregon Museum of Science and Industry in the United States highlights the importance of this principle of reciprocity, among many other interesting insights and recommendations when working inclusively with Native American nations and other diverse community partnerships (Coats et al, 2006; OMSI, 2015).

Conclusion

Inclusion is the art of intentionally designing learning experiences where people from every given background, ability, culture, language, socio-economic status, and level of education, among other intersectionalities (Crenshaw, 1989), can participate on a par with the mainstream audience. Shelley Moore (2016) describes it as a process and a way to assess where an institution is in its journey to inclusion. She explains that this process is also connected to the history of each community, which requires that institutions undertake in-depth community-based understandings of their localised publics. Institutions move progressively from exclusion to segregation, from segregation to integration, and from integration to inclusion to a state where there is no 'us' and 'them', only 'we' (Moore, 2016). Inclusion, Moore (2019) explains, is about identity and diversity, a celebration of our uniqueness and strengths, where all characteristics or intersectionalities are valued. Inclusion 'is looking out for who is missing', and it is the job of educators and science communicators to figure out the missing intersections and develop inclusive landscapes that will bring them on board.

The road to becoming an inclusive science communication institution is a challenging but worthwhile endeavour that requires a lot of self-reflection, critical institutional analysis, and the identification of institutional strengths, capacities, and limitations. It also involves an evaluation of institutional best practices that are already in place, as well as those that are not, with strategic planning to enhance existing good practice infrastructure. It is critical in this realm to consider key milestones such as adapting to different audiences as a great start, building community partnerships and incorporating their cultural wealth, knowledge, and goodwill (Yosso, 2005). In addition, collaborative partnerships and networking with other science communication institutions that are on a similar journey to inclusion provides myriad opportunities to engender, upscale, and sustain solidarity, cooperation, transferrable skills, and knowledge sharing, as well as inspiration and encouragement, to sustain momentum and build on achievements and progressive developments across the board.

References

Coats, V., Maryboy, N., Begay, D. (2016) 'Roots of wisdom. Native knowledge. Shared science. And collaboration with integrity', *Exhibition*, [online] Fall: 50–59, Available from: https://static1.squarespace.com/static/58fa260a725e25c4f30020f3/t/59f7f4f627ef2d4ad32e0fe2/1509422328463/12_RootsOfWisdom.pdf

Crenshaw, K. (1989) 'Demarginalizing the intersection of race and sex: a Black feminist critique of antidiscrimination doctrine, feminist theory and antiracist politics', *University of Chicago Legal Forum*, 1: 139–67, Available from: https://chicagounbound.uchicago.edu/cgi/viewcontent.cgi?article=1052&context=uclf

Dawson, E. (2013) 'What is science communication?', Catalyst, October, pp 4–5, Available from: https://www.stem.org.uk/system/files/elibrary-resources/legacy_files_migrated/29743-Catalyst%2024%201%20556.pdf

Eid, N. (2016) 'Integration vs inclusion, the role of ICT accessibility in education system', LinkedIn, 15 March, Available from: https://www.linkedin.com/pulse/integration-vs-inclusion-role-ict-accessibility-education-nabil-eid/

Leyman Pino, A., Redmond-Jones, B., and Wawerchak, V. (2020) 'Blended language programming', *Informal Learning Review*, 159: 7–13.

Monterrey Bay Aquarium (2019) '2019 guest and community highlights', Available from: https://www.montereybayaquarium.org/globalassets/mba/pdf/guest-highlights/monterey-bay-aquarium-guest-community-highlights-2019.pdf

Moore, S. (2016) *One without the Other*, Winnipeg, Canada: Portage and Main Press.

Moore, S. (2019) 'Cranbook', Available from: https://blogsomemoore.files.wordpress.com/2019/06/cranbrook.pdf

National Disability Authority (2020a) 'The 7 principles', Available from: https://universaldesign.ie/what-is-universal-design/the-7-principles/the-7-principles.html

National Disability Authority (2020b) 'What is Universal Design', Available from: https://universaldesign.ie/what-is-universal-design/

Oregon Museum of Science and Industry (OMSI) (2015) 'Roots of Wisdom, Native knowledge, shared science: reflections and ideas about collaboration with integrity', Available from: https://static1.squarespace.com/static/58fa260a725e25c4f30020f3/t/59f7f4f627ef2d4ad32e0fe2/1509422328463/12_RootsOfWisdom.pdf

UK Office for National Statistics (ONS) (2023) 'How life has changed in Leicester: Census 2021', Available from: https://www.ons.gov.uk/visualisations/censusareachanges/E06000016/

United States Census Bureau (2019) 'Quick facts California', Available from: https://www.census.gov/quickfacts/CA

Vinter, R. (2023) '"Diversity is a beautiful thing": the view from Leicester and Birmingham', *The Guardian*, [online] 29 November, Available from: https://www.theguardian.com/uk-news/2022/nov/29/leicester-birmingham-first-super-diverse-uk-cities-census

Yosso, T.J. (2005) 'Whose culture has capital? A critical race theory discussion of community cultural wealth', *Race Ethnicity and Education*, 8(1): 69–91, Available from: https://thrive.arizona.edu/sites/default/files/Whose%20culture%20has%20capital_A%20critical%20race%20theory%20discussion%20of%20community%20cultural%20wealth_1.pdf. doi: 10.1080/1361332052000341006.

4

Building Capacity for Science Communication in South Africa: Afrocentric Perspectives from Mathematical Scientists

Mpfareleni Rejoyce Gavhi-Molefe and Rudzani Nemutudi

Introduction

Science communication and transformation policy frameworks in South Africa

In South Africa, the proliferation of the multifaceted science communication agenda has seen calls for action from the national government for scientists and researchers to demonstrate the broader societal impact of their research. The impact involves bridging the longstanding gap between science and South African society through a myriad of platforms, tools, and formats as reflected in the National Research Foundation (NRF) Strategy 2025 (NRF, 2020), National Development Plan (NDP): Vision 2030 (NDP, 2012), and the Department of Science and Technology (renamed Department of Science and Innovation [DSI] in 2019) policies and strategies (DSI 1996, 2015, 2019). It was envisaged that closing the gap between science and society would help build a transformative and effective National System of Innovation (NSI) for the country's sustainable development, progressive democracy, and the growth of a literate, informed society. The 1996 white paper on science and technology, 'Preparing for the 21st century' (DSI, 1996), had long provided the overarching science, technology, and innovation (STI) vision for South Africa in the post-apartheid era.[1] The white paper demonstrated the national government's commitment to supporting and advancing the science communication agenda in South Africa after the country became a democracy in 1994. The policy document further paved the way for policy and programmatic frameworks for science and society engagement in South Africa. To begin this process, in 1998 the first national series of public science events was launched as part of a year-long programme – 'Year of Science and Technology' (YEAST). The programme was launched by the DSI (then the Department of Arts, Culture, Science, and Technology). YEAST was celebrated through various science communication platforms. It became an annual event that rapidly grew in size and scope across the country. Subsequently, the science promotion directorate was established, and various strategies, interventions, and institutions at the government level to support the science communication agenda have been developed. Resources have been allocated towards these priorities. However, these development plans were enacted without a coordinated national strategic framework at the policy, implementation, and programmatic delivery levels.

The adoption of the National Science Engagement Strategy (SES) in 2015, followed by its Implementation Plan in 2017 and its 2020 Monitoring and Evaluation frameworks, marked a milestone in advancing and consolidating the thrust of the science communication agenda in the country (DSI, 2015, 2017, 2020). The SES stipulates that scientists are one of the critical actors in achieving its objectives. It also urges scientists to use science communication activities as a platform to instil a culture of engaging the public on science issues and 'systematise science outreach and/or science engagement activities in alignment with the level of research resources allocated to them' (DSI, 2015, p 24). The SES Implementation Plan specifically enumerates eleven target audiences/publics for science communication in South Africa. These publics include: learners, educators, students, scientists and researchers, science interpreters, industry, decision-makers, journalists, tourists, Indigenous knowledge holders, and the general public (DSI, 2017, p 12). The 2019 white paper on STI, designed to address the gaps in the 1996 version, reaffirmed a call for scientists to engage actively with the broader society (DSI, 2019). Undoubtedly, South African government policy imperatives for science communication are comprehensive and compare favourably with similar policies elsewhere in the world. However, considerable gaps exist between these comprehensive policy frameworks and the delivery of effective, meaningful, and impactful science communication and public engagement practices by South African scientists and researchers.

At the level of the African continent, science communication has a similarly critical role, particularly when viewed against the backdrop of persisting structural and infrastructural deficiencies, coupled with poor public scientific literacy that continues to hamper the overall STI developmental plan of the continent. Thus, without advancing the science communication agenda, attempts at addressing pressing socio-economic development and transformation challenges while attempting simultaneously to mould a scientifically literate African society are doomed to fail.

Science communication and pan-African policy frameworks in Africa

The African Union (AU) Agenda 2063, 'The Africa we want', and its STI strategy for Africa (STISA 2024), which are firmly rooted in the ideals of pan-Africanism and the African Renaissance, provide transformative frameworks for addressing the continent's peculiar challenges and place STI at the centre of Africa's developmental agenda (AU, 2014, 2015). STISA 2024 affirms the AU's commitment to the science communication agenda, as evident in the adoption of Priority 3, 'Theme on Communication (Physical and Intellectual Mobility)', in its strategy. However, Priority 3 details a very limited framework and direction on how Africa will grow the footprint of science communication on the continent and hence accelerate the growth of STI. Such dynamic pathways are critical at the continental level because many African countries still lack substantive strategic development in STI (AU, 2019). Many of those countries with STI strategies often do not have concrete plans on growing Africa's capacity for the science communication agenda. A concrete plan for the science communication agenda on STISA 2024 as a top-down approach will ensure greater alignment between national and sub-regional policies to enhance coordination in strategic approaches to science and society engagement. The recent five-year Implementation Report of STISA 2024 (AU, 2019) recognises the importance of science communication on the continent in Priority 5, 'Live Together – Build the Society'. The STISA report further points to the commitment of the South African government to science communication. However, no clarity is provided on how African governments should drive the science communication agenda with the relevant stakeholders as critical partners, and most importantly, how adequate resourcing models are to be implemented to build the human capacity for this agenda across the continent.

Despite the limitations of STISA 2024 in creating an enabling environment for the development of science communication human resource capacity and mechanisms on the continent, the footprint of the science communication field and practice has continued to grow, albeit slowly. This is due to the investment in the STI science communication agenda in some countries and the growing leadership of some African stakeholder institutions working for the advancement of science and society engagement on the continent. The Lagos Declaration and Call to Action on Science Communication and the Public Learning and Understanding of Science in Africa (2016) – a seminal framework and the first of its kind in Africa – is one example. It was adopted by multiple African stakeholders in 2016 to address the limitations of STISA 2024 and provide concrete steps, enablers, and interventions at the policy, institutional, and individual (for example, scientists) levels (African Gong, 2016). The framework calls on and encourages African scientists and scientific institutions to mobilise action, policy developments, programmes, and capacity building for the delivery of a transformative, multidisciplinary, inclusive, and empowering Afrocentric science communication agenda for the transformation of the quality of life and well-being of the continent's citizens (Rasekoala and African Gong, 2019).

Science communication in South Africa: overcoming historical legacies of race, gender, and socio-economic inequalities

Like other African nations, South Africa continues striving to overcome the pernicious legacies of the colonial era. However, South Africa's societal scars have been deepened by the brutal system of apartheid that has effectively deprived Black South African and female citizens from equitable access to science, technology, engineering, and mathematics (STEM) education, skills, professions, and economic resources.[2] The 2018 National Advisory Council on Innovation's discussion document alleges that the transition to post-apartheid South Africa has resulted in some positive progress (see Maharajh, 2018). Furthermore, efforts are being made to build an inclusive, scientifically literate society to make South Africa one of the world's scientifically and technologically advanced countries (DSI, 2015). With Black African females constituting the majority of the population, it has been increasingly important for the government to address the racial and gender imbalances prevalent in the STI system over previous decades.

South African science policy makers perceive science communication as a tool to redress the lasting impact and legacies of apartheid related to socio-economic, cultural, racial, and gender social justice inequalities within the STI space, and to transform the higher education system (DSI, 1996; Joubert and Mkansi, 2020). South Africa was the first African country to strengthen and transform its science–society agenda by developing a Science Engagement Strategy (DSI, 2015). One of the advantages of South Africa is that its STI system has built within it the science funding councils such as the NRF. Similar public science granting councils exist in East, West, and Southern Africa and are coordinated by the Science Granting Councils Initiative in sub-Saharan Africa. South Africa has a relative advantage in its potential to implement a dynamic science communication agenda. Government and broader stakeholder participation is aided by well-resourced implementation models. However, despite the relative proliferation of structural and infrastructural resources in South Africa, the gap between the well-developed science policy frameworks and their implementation remains.

One of the problems in resolving this issue is that there is a lack of essential foot soldiers and resources to drive the country's effective and sustainable delivery of science communication activities. Given the urgent need to equip and grow the critical mass of African science communicators within holistic, sociocultural, and empowering ethical frameworks, the challenge is daunting. This is not surprising given that the South African STI ecosystem is yet to be transformed in terms of race, gender, and social class demographics despite the established national government's policy frameworks, initiatives, and mechanisms (see Joubert, 2017). This profound need arises amid the reality that longstanding colonial and Eurocentric hegemonic traditions still dominate STEM education, research, and communication in Africa. Such traditions continue to undermine and relegate to the margins African ethics, languages, values, Indigenous knowledge systems, cultures, practices, and contributions to scientific advancement globally, resulting in stagnation, sociocultural exclusion, and systemic structural inequalities (Rasekoala, 2020a, 2022; Kago and Cissé, 2022). Furthermore, the need also arises amid the reality that not only is the science communication field itself still at a very nascent/emerging stage on the African continent but the actors in the field on the African continent are predominantly male as opposed to elsewhere in the world (Rasekoala, 2019).

The number of African women participating in STI remains low across the continent. In South Africa, the underrepresentation of women in the STI system is further exacerbated by the history of colonialism and apartheid, which portrayed Black African women as minors and, in recent times, as outsiders within academia (Idahosa and Mkhize, 2021). According to the South African NRF Information Portal (see NRF, 2021), analysis of five NRF rating categories across race and gender (last updated on 31 August 2021) reveals the following:

- Gender Analysis: Only 35 per cent of rated South African researchers in 2021 are women (although they account for 51.1 per cent of the country's total population).
- Racial Analysis:
  - 66 per cent of rated researchers are White African (despite making up only 7.8 per cent of the country's total population).
  - 21 per cent are Black African (even though they make up the majority of South Africa's total population [80.2 per cent]).
  - 8 per cent are Indian/Asian (they make up 2.6 per cent of the country's population).
  - 5 per cent are Coloured[1] (the people of dual heritage, who represent 8.8 per cent of the total population).
- Race and Gender Intersectionality Analysis:
  - 39 per cent of rated White African researchers in South Africa are males, while 27 per cent are females of the same race.

- 17 per cent of rated Black African researchers are males, while only 4 per cent are females of the same race.
- 5 per cent of the rated Indian/Asian researchers are males, while 3 per cent are females of the same race.
- 3 per cent of the rated Coloured researchers are males, while 2 per cent are females of the same race.

The NRF rating serves as a benchmarking system of the quality of South African researchers against the best in the world. Individuals are rated based on the quality and impact of their research outputs over the past eight years. The findings highlight the profound under-representation of Black African researchers in the national STI system, especially women. This is of grave concern because South Africa's societal transformation, its STI development trajectory, and science communication agendas are highly likely to fail without addressing these pernicious inequalities of race and gender in its STEM human capital development profiles.

This chapter argues that critical gaps exist between African governments' policy positions and the practical implementation of scientists' imperative to deliver, through meaningfully resourced mechanisms, a viable, sustainable, and impactful science communication agenda. As a dynamic way forward, we propose the adoption of impactful African-focused transformative initiatives, such as the innovative Afrocentric capacity-building programme entitled 'Africa Scientifique: Leadership, Knowledge & Skills for Science Communication'. We further demonstrate how this initiative can be used to address several of the aforementioned gaps and challenges in the South African and African continental science communication landscape. The Africa Scientifique (AS) programme – a case study for the present chapter – has been delivered to three cohorts of master's students at the African Institute for Mathematical Sciences (AIMS) in Muizenberg, South Africa, in 2020, 2021, and 2022. As a point of departure, we also discuss some critical issues affecting the growth of the science communication agenda in Africa.

Challenges of growing the critical mass for science communication in Africa

A critical challenge to growing the African science communication human resource capacity is the lack of training interventions and the dearth of adequately resourced capacity-building opportunities across the STI ecosystem. While this is not a unique challenge in developing countries, it is worse within the African context, and it has significantly contributed to African scientists' poor skills, knowledge, motivation, and expertise in science communication (Karikari et al, 2016; Ndlovu et al, 2016; Rasekoala, 2020b; Gavhi-Molefe et al, 2021). Science communication training programmes have been established and normalised to a large extent in many countries across the Global North (Silva and Bultitude, 2009; Heslop et al, 2021). However, very few training opportunities to equip scientists and researchers with these necessary and relevant skills exist on the African continent, despite the commitments of African governments to the science communication agenda. Moreover, even within the South African context, with its relatively robust, comprehensive STI and science engagement policies and strategies, the limited existing science communication training programmes remain fragmented, without inclusive, overarching, and coordinated frameworks. Thus, in the end, the developed frameworks are rendered less impactful locally and almost inconsequential continentally.

Furthermore, African science and its communication landscape are already dominated by Eurocentric hegemonic traditions that have often served to entrench the general interests and preferences of the Global North over those of the South. This is also evident in some African science engagement programmes, which are not only one-off unsustainable events but adopt models that are customised to the agendas of Western donors at the expense of local realities (Kago and Cissé, 2022). These traditions and agendas assume that African scientists cannot be key players and creators of innovative and transformative knowledge and successful science communication practices (Rasekoala, 2020b; Finlay et al, 2021; Rasekoala, 2022).

Case study: Afrocentric science communication training and capacity building

The science communication agenda has always been integral to the AIMS vision and mission, as stated in its Institutional Strategy (AIMS, 2021). AIMS is a pan-African network of affiliated institutes dedicated to postgraduate education, research, and public engagement in the mathematical sciences. AIMS centres are currently located in South Africa, Senegal, Ghana, Cameroon, and Rwanda. In addition to conducting cutting-edge research, AIMS students and scientists are expected to engage interactively with wider audiences. However, it gradually became evident that science communication skills and aptitude were a missing piece in the puzzle among AIMS students and scientists. AIMS students and scientists required skills to undertake mathematics outreach activities effectively. In 2020, as one of the strategic interventions to address these identified needs, AIMS partnered with African Gong – the pan-African network for the popularisation of science and technology and science communication – to deliver the AS programme, which addressed the science communication skills gap. AS is a unique, innovative Afrocentric capacity-building programme conceptualised to support young and emerging African scientists in order to advance research expertise and science communication capacity on the African continent.[3]

It is designed as a highly practical, hands-on minds-on, intensive, inclusive, transformative, and interactive three-phased capacity-building framework. It entails the following building blocks:

- a short introductory workshop where participants are introduced to the background, transformational Afrocentric concepts, the rationale behind the AS programme, and the envisaged outcomes and benefits for participants;
- an intensive three-day workshop where participants delve deeper into the realms of science communication, science and society, public engagement, and the potential leverages that these skills, activities, and practices can engender in their research and career advancement; and
- a six-month post-workshop period where participants have direct, peer-to-peer, team mentoring, and facilitated support while planning, delivering, monitoring, and evaluating their public engagement activities over the six months' timeline.

In the remaining part of this chapter, the authors reflect on the AS programme with a particular focus on the three-day phase 2 workshops that were delivered in 2020, 2021, and 2022 to the cohorts of master's students at the AIMS South Africa Campus in Muizenberg, Cape Town. The participants – 15 in 2020 (26 per cent of whom were female), 25 in 2021 (48 per cent of whom were female), and 30 in 2022 (48 per cent of whom were female) – were aspiring mathematicians, statisticians, physicists, and computer scientists from four African countries in 2020 (South Africa, Cameroon, Nigeria, and Zambia), seven African countries in 2021 (Democratic Republic of the Congo, Kenya, Madagascar, South Africa, Swaziland, Zambia, and Zimbabwe), and nine African countries in 2022 (Malawi, Nigeria, Rwanda, Cameroon, Madagascar, South Africa, Botswana, Zambia, and Zimbabwe). During the introductory workshops, all three cohorts (2020, 2021, and 2022 groups) indicated that they had never been afforded the opportunity to participate in any science communication capacity-building training nor undertaken any public engagement activity to communicate their mathematics to the public. This was not surprising given the limited opportunities for science communication training in the South African science ecosystem and across the African continent. It should also be noted that there is a perception among the mathematical sciences community that it is particularly challenging to communicate mathematics. For these students, the AS programme was thus the beginning of a new journey of knowledge, discovery, and first-hand experience in science communication and, therefore, a potential game changer. The following sections provide some insights into the programme outcomes for the participants and how they fit within the three unique building blocks of the AS programme, namely Afrocentricity; transformation in terms of skills, mind-sets, and worldviews; and the learning by doing methodology.

AS programme: Afrocentricity as a guiding philosophy

The pan-African focus that underscores the partnership between AIMS and African Gong forms the basis for adopting Afrocentricity – the guiding philosophy around which the AS programme has been conceptualised, designed, planned, organised, and delivered. In his seminal work titled *The Afrocentric Idea*, Asante (1987, p 2) defines Afrocentricity as an act that 'literally places African ideals at the centre of the analysis (in every aspect) that involves African culture and behaviour'. Thus, if African authenticity is to be restored and reclaimed, Afrocentricity has to permeate all areas of studies where people of African descent are involved. This should include the shaping of modes of communication that mediate the transmission of knowledge between the experts and the public in fields such as STEM. In essence, Afrocentricity is a philosophy that focuses on the challenges confronting Africans across the African continent. With an Afrocentric outlook as their guideline, Africans can afford themselves time and space to formulate homegrown solutions informed by their geographical locations and cultural backgrounds.

The thrust and underlying objectives of the AS programme are to provide training and mentorship to AIMS master's students from across the African continent. The students were equipped with tools to become well-rounded scientists who can devise and drive the Afrocentric STI ecosystem and science communication agenda on the continent and set it on a pathway to liberate it from Eurocentric dominance (Finlay et al, 2021). Hence the overarching theme of the programme was aptly titled 'Growing Africa's Capacity for Science Communication Impact'. The workshops were ably facilitated by dynamic and diverse African science communication experts and practitioners from academia, science councils, research institutions, and industry. The associated sessions covered a broad spectrum of specialised thematic topics tailored to the needs of the participants. The workshops also provided a unique opportunity for participants to directly engage in dialogue with key South African government science policy makers. The inclusion of the policy makers also paved a way towards addressing the need for African governments and institutions to work together in strategic partnerships to deliver homegrown Afrocentric, sustainable, transformative, and inclusive science communication-based capacity-building programmes on the continent. The 2021 and 2022 workshops also provided a unique opportunity for participants to interact with some of the 2020 and 2021 AS programme alumni pursuing careers in academia and industry. These AS alumni also contributed to the workshops. This element of the AS programme provides participants with the opportunities to learn from and be inspired by AS alumni cyclic learning and encourages the AS alumni to continue their science communication/public engagement good practices.

Post-workshop testimonials and reflections were collected from the participating students and the session facilitators/trainers. The facilitators' feedback was enriching and instructive in leveraging wider resources and structural support to broaden the participation in science engagement roll-out and implementation plans. In addition, the student testimonials echoed the transformational impact of the workshops.

AS programme: transformation for sustainable impact

Transformation agendas and plans are necessary tools through which society can address the socio-economic challenges and inequalities it is confronted with. The transformational impact of such plans is often more meaningful if the plans themselves are designed to be long-lasting and transgenerational. The impact can best be effected through implementations that are targeted at a relatively younger generational mix with an emerging career line of sight still spanning a longer time horizon ahead of it. The AS workshops were strategically designed with medium- to long-term transformational impact, targeting student participants whose average age group ranged from 20 to 30.

The students offered heart-warming testimonies of how the workshops had profoundly challenged and assisted them in recognising and confronting, head-on, their communication blind spots, capability gaps, and pitfalls. Several students highlighted the experience of facing the audience to communicate, which was an important aspect of the workshops. The experience presented them with a less intimidating platform from which they could build on their presentation skills and become more confident public speakers. The workshops' format pushed the students out of their comfort zones. They were expected to participate and make personal inputs during discussions in groups and to speak in front of their participating peers about one aspect or another discussed during the group sessions, including their own research projects and sociocultural inclusion/exclusion experiences. The discussions were particularly important. Incorporating discussion and reflection sessions can prove to be the best practice in STEM communication training, as it 'encourages students/trainees to be responsible for their own learning' (Silva and Bultitude, 2009, p 11).

During the sessions of the three workshops, the more reserved students became more interactive with each other and the facilitators as the sessions progressed. The student testimonials also highlighted how the workshops' emphasis on societal benefits/impacts challenged and helped the students reaffirm their moral and ethical obligation to society as aspiring scientists. They appreciated that "many more people out there need to be aware of the potential or realised benefits of the scientific research done by scientists or by students in their laboratories or out in the fields", which makes communication even more relevant as a strategic skill. The workshops included a recognition award for excellence for the best female and male students (two gender-based awards) who had demonstrated outstanding improvements in their engagement, enrolment, communication, and presentation skills over the course of the three days of the workshops. One of the award winners for the 2021 workshop remarked:

> 'After attending the workshop, one is more aware that it is good and even better to do research in mathematics and in the sciences, but it is, even more, better to be able to share this knowledge with others in the community, because many people need to be aware of what research is being done, and how they can benefit from it.'

Issues of gender equality in STEM and the impact of the workshop on overcoming the challenges of patriarchal attitudes in academia and society were also highlighted: "Being a woman is hard, we still have males who think women are supposed to do certain things. Growing up I always believed that women are not supposed to stand and give a talk in front of men" (AS 2020 workshop participant).

Another female participant gave the following feedback:

> 'I have been empowered by the Africa Scientifique programme to learn that I am enough as a female scientist! I believe that I can do it! – I can make the change that Africa really needs to bridge the gap that has been there for so long in the STEM fields'. (AS 2022 workshop participant)

English was the medium of communication throughout every phase of the AS programme. Nevertheless, there was a strong emphasis on the importance of using Indigenous African languages to demonstrate language plurality and inclusivity in communicating science. After all, such an initiative, as a vehicle for promoting science and its implications in everyday lives, has the potential to increase trust and participation in science across different publics in Africa (Kago and Cissé, 2022). In addition, in some of the sessions, the participants were actively encouraged and enabled to engage in exploring the empowering impact of sociocultural inclusion in science communication using the medium of their local African languages and Indigenous knowledge systems. This critical element of the AS programme provides participants with a basis for featuring Indigenous African languages in conceptualising and delivering their post-workshop science communication projects to reach different audiences, particularly groups of hard-to-reach and/or neglected and marginalised communities.

The workshop's goals and objectives received an overwhelmingly positive reception because they resonated with the students' expectations and went a long way to address the identified skill gaps.[4] At the end of the workshop, most students embraced the notion that a successful and effective science communication strategy does not require a communication specialist. The students simply have to be comfortable giving presentations to different public audiences on how their research projects can solve societal challenges. The outcomes of their projects need to be communicated in a message packed with captivating tools and language that is easy to understand, without employing heavy usage of technical jargon and over-specialised terminology. Moreover, the students realised that to be more confident in communicating science and become visible and successful scientists, it is essential to keep upskilling themselves, practice, learn from other well-rounded scientists (or role models), and learn by doing.

AS programme: learning by doing with peers and facilitated support

The third unique element of the AS programme challenges the students to learn by doing with their peers and facilitated support. This approach is critical to sustaining engagement practice beyond the workshop training. In essence, the approach adopts the proverb, 'Give a man a fish, and he will eat for a day. Teach a man how to fish, and you feed him for a lifetime'. This approach allowed participating students to think of themselves as scientists capable of harnessing their imaginations, innovative capacity, and creativity and using their mathematical knowledge, analytical capabilities, and skills to develop their science communication practices. Over the three days of the workshops, all sessions were designed to incorporate hands-on minds-on activities to urge the students to engage actively among themselves in small groups and with facilitators. Each student was given at least one opportunity during each day of the three-day workshop to present the outcome of a group discussion on a range of topics covered within a dedicated session with support from peers and the facilitators. All the sessions ended with a reflective activity to engender insightful understandings and transformative perspectives that motivated behavioural and mind-set changes. Group activities were tailored to meet the principles of inclusivity, equity, and diversity in terms of gender mix and geographical spread. This provided participants with first-hand experience of how gender and sociocultural inclusion challenges in science communication can be addressed.

On the last day of the workshops, the participants were offered an opportunity to conceptualise, design, and present proposals for their science communication projects' practical activities in three minutes using one PowerPoint slide with coaching, guidance, and constructive feedback from the facilitators. The intention was to encourage them to incorporate their learning outcomes into the actual proposal plans of the science communication projects. Emphasis was placed on the importance of identifying potential project outcomes for societal benefits as the basis for a sound and impactful public communication plan. Thus, the learning by doing approach was practical and compelling. Most students appreciated and highly valued the exercises, as they allowed them to showcase their creativity and presentation skills. Indeed, as one student participant succinctly put it: "Getting a chance to conceptualise, prepare and present our proposals for mathematical science communication projects with social impact made me realise the potential solutions we have as African scientists for our communities."

Similarly, most students described the learner-centred, group work, peer-to-peer support, and collaborative framework of the AS workshops as a game changer. The approach provided a springboard for sharing ideas and innovative solutions designed to inform comprehensive science communication plans that mirror everyday experiences. Indeed, one participant's testimonial at the end of the workshop alluded to this: "Working in groups with peers is an important thing because we can share ideas and come up with good solutions to solve the sociocultural issues facing our African countries."

Most importantly, in a deliberate drive to incorporate into the participants' mind-sets the Afrocentric conceptual framework that underpinned the AS programme, the participating students' science communication project proposals needed to feature diverse target audiences, tools, formats, mathematical science themes or concepts, use of African Indigenous languages, and address societal issues in their communities as the basis for a comprehensive, impactful, and transformative science communication activity plan. This had encouraging results when the students presented their individual science communication project plans during the last session on the final day of the workshops:

- Participants were able to identify a range of hard-to-reach and/or neglected audiences defined in the DSI Science Engagement Implementation Plan (DSI, 2017). Examples of such audiences included township taxi managers and drivers, social grants holders, social media users, small-scale farmers, Indigenous knowledge holders, traditional healers, shopkeepers, township youth, elderly patients, tourists, and government officials, all of whom are critical components of an inclusive African communal and societal setting. These audiences go beyond the usual convenient targets of school learners and educators that most science communication activities in South Africa and on the African continent are usually designed for and directed at.
- Participants were able to articulate the African Indigenous languages such as Setswana, Tsonga, Swahili, isiZulu, Tshivenda, and Luhya that

they will use to deliver their science communication activities. This was an important approach that was emphasised during the workshop. After all, communicating science in African Indigenous languages remains one of the critical challenges across the continent (Ramos and Empinotti, 2017). African languages inclusion was one of the most important tools and underlying concepts that were emphasised during the workshop.

- Participants conceptualised various science communication projects that addressed societal challenges. They planned to engage with the publics on the use of mathematical science knowledge and research work, such as: enhancing COVID-19 vaccination strategies; the dangers of social media; corruption in governments; starting a small business; curbing drug abuse among the youth; minimising traffic jams and accidents; problem gambling; drink driving; and the best routes to minimise township taxi journeys using mathematics and mathematical models.

Conceptualising these diverse science communication projects was a demonstrable impact of the workshops on the participants' horizons, empowerment, and transformation. The participants were now science communicators grounded in their African indigeneity. As one of the participants remarked: "Mathematics is a key solution to the existing world problems, but we emerging African scientists need to be able to share it in the communities' local languages so that it impacts many people."

Conclusion

There is an urgent need to grow the science communication human resource capacity in South Africa and across the African continent. Doing so is critical because growing such capacity could be a spur for driving the science communication agenda from an Afrocentric-inspired point of view. African communities should be the main focus, targets, and agenda-setting nodes for the conceptualisation and realisation of this transformative capacity-building vision to grow the science communication footprint across the African continent. As a point of departure in this journey, African governments need to take a pragmatic approach towards science communication and its strategic expansion across the continent. There is a profound dichotomy bedevilling the science communication agenda on the continent. On one hand, the policy and strategic imperatives of governments point to the need for African scientists to demonstrate the relevance of their research to society, with concomitant emphasis on communicating and engaging directly with public audiences. On the other hand, the requisite resources and training opportunities are not made available by the same governments to equip, enable, and support the African science communication drivers with fundamental tools and confidence-inspiring infrastructure to deliver sustainable and impactful science communication activities. This is a critical challenge that one of the AS 2021 workshop facilitators eloquently articulated in their testimonial:

> The main challenge is that policymakers together with those who have authority to direct the distribution of resources do not seem to regard science communication as an essential and strategic component of the operational portfolios of any publicly funded institution, especially those that are focused on research and its impact on society. Science communication has to be elevated to the same level as any other portfolio of a publicly funded institution in order for society to appreciate both the return on, and the value of their public investment.

African governments, universities, and research institutions need to work together in strategic partnerships to deliver Afrocentric, sustainable, transformative, and inclusive science communication-based capacity-building programmes on the continent to allow sharing of resources and best practices (Rasekoala and African Gong, 2019). Although structural changes are underway in Africa, with developmental frameworks such as AU's Agenda 2063 to build 'the Africa we want', the pace of progress is not yet at the desired levels. Thus, for the continent to grow the science communication transformation agenda (which has been hampered by sociopolitical factors), its project of growing the scientifically literate African citizenry should be fully accelerated. This can only happen if African scientists are equipped with the necessary skills, infrastructure support, and resources.

The AS programme demonstrates that while African practitioners can learn practices from elsewhere, they need to conceptualise and support Afrocentric capacity building for African scientists and researchers to liberate the STI ecosystem and science communication agenda from longstanding Eurocentric dominance. With a greater commitment and investment in training interventions such as the AS programme and adequately resourced capacity-building opportunities across the science ecosystem, African science communication human resource capacity can flourish and grow within the empowering paradigm of a renewed African Renaissance. African governments' support and recognition of the few existing sustainable and impactful training programmes could be the first step in closing the longstanding gap between their science policy imperatives and science communication practice. It has been shown that training is one way scientists can gain the mind-sets and skills needed for effective science communication and engagement (Heslop et al, 2021). Indeed, the impact of mind-set changes in the participants in the AS programme elucidated many ringing endorsement of public engagement and science and society interactions in

Africa. The following is an exemplar testimonial from a participant of the 2020 workshop:

> Public engagement activities enhance the use and the discovery of science by other people. Public engagement is the light which we have to use to bring science to other people. Public engagement is a key step to disseminate our research. Public engagement is like an open door between scientists and society. Public engagement is a kind of communication about our scientific competence.

In summary, the observations, outcomes, and experiences gathered from the three AS workshops validate the efficacy of interactive training methodologies as viable tools that relevant institutions could deploy to develop critical human resource capacity to drive the science communication agenda across the African continent. Universities and research institutions offering graduate education in STEM need to consider elevating science communication as a critical skill that should form an integral part of a research and training development curriculum. If incorporated at the early stages of the students' training pathway, foundational science communication skills would go a long way to counteract gaps in communication know-how, motivation, and aptitudes. It has been shown that science communication actions, practices, and drive are dynamic transformational tools for post-AIMS career advancement and progression across critical parameters.

Government policy makers whose decisions influence the allocation of resources need to be made aware of the imperatives of science communication as an essential skill for which training can be offered to graduate students and early career researchers. Capacity-building interventions such as the AS programme have the potential to close science communication gaps. Furthermore, these programmes would go a long way towards achieving critical science communication delivery and implementation milestones when planned from and implemented within an Afrocentric perspective.

Notes

[1] Apartheid (translated from the Afrikaans or West Germanic language meaning 'apartness') was a racial policy that governed relations between South Africa's ruling White minority and its Black majority for much of the latter half of the 20th century; the system sanctioned racial segregation and political and economic discrimination against non-Whites (Britannica, 2023).

[2] South African socio-demographics and cultural characteristics:

- The current population of South Africa is 60.14 million people, around 51.1 per cent (30.75 million) of whom are women (Department of Statistics South Africa: Stats SA, 2021).
  The country is a multiracial society with four racial/ethnic population groups (using standard South African categories by law), namely Black African, Coloured (people of dual heritage), Indian/Asian, and White.

- According to Stats SA (2021), Black Africans constitute the majority of the population (80.2 per cent), followed by Coloureds at 8.8 per cent, then Whites (7.8 per cent), and finally Indians/Asians (2.6 per cent).
  About 28.3 per cent of the population is aged younger than 15 years (17.04 million) and around 9.2 per cent (5.51 million) is 60 years or older.
- South Africa is a multilingual society with 11 official languages, namely: English, Afrikaans, isiNdebele, isiXhosa, isiZulu, Sesotho, Setswana, Tshivenda, Sepedi, siSwati, and Xitsonga.
  It should be noted that there are an estimated 2,000 languages spoken on the African continent (Kago and Cissé, 2022).

[3] Africa Scientifique Programme: https://www.africangong.org/capacity-building-flagship-programmes/

[4] AIMS House of Science YouTube Channel. For videos of AS participants sharing their workshop experiences, outcomes, and benefits, see: https://www.youtube.com/channel/UCk54pq7mCovkGz-GkptdKpQ

References

African Gong (2016) 'Lagos declaration and call to action on science communication and PLUS in Africa', Lagos State Ministry of Health, Available from: http://www.africangong.org/wp-content/uploads/2019/03/lagos-declaration-and-call-to-action.pdf

African Institute for Mathematical Sciences (2021) 'AIMS 2021–2026 strategic framework', unpublished.

African Union Commission (2014) 'Science, technology and innovation strategy for Africa 2024: Addis Ababa, Ethiopia; African Union Commission', Available from: https://au.int/sites/default/files/newsevents/workingdocuments/33178-wd-stisa-english_-_final.pdf

African Union Commission (2015) 'Agenda 2063: the Africa we want', Available from: https://au.int/sites/default/files/documents/36204-doc-agenda2063_popular_version_en.pdf

African Union Commission (2019) 'Third ordinary session for the specialized technical committee on education, science and technology (STC-EST) 10th to 12th December 2019, Addis Ababa, Ethiopia: contextualising STISA-2024; Africa's STI implementation report 2014–2019', Available from: https://au.int/sites/default/files/newsevents/workingdocuments/37841-wd-stisa-2024_report_en.pdf

Asante, M.K. (1998) *The Afrocentric Idea*, Philadelphia, PA: Temple University Press.

Britannica, T. (2023) *Apartheid. Encyclopedia Britannica*, Available from: https://www.britannica.com/topic/apartheid

Finlay, S.M., Raman, S., Rasekoala, E., Mignan, V., Dawson, E., Neeley, L. et al (2021) 'From the margins to the mainstream: deconstructing science communication as a White, Western paradigm', *Journal of Science Communication*, 20(1): C02.

Gavhi-Molefe, M.R., Jensen, E., and Joubert, M. (2021) 'Why scientists agree to participate in science festivals: evidence from South Africa', *International Journal of Science Education, Part B*, 11(2): 127–42.

Heslop, C., Dudo, A., and Besley, J. (2021) 'Landscape of the UK science engagement training community', Center for Media Engagement, Available from: https://mediaengagement.org/wp-content/uploads/2021/09/Landscape-of-the-UK-Science-Engagement-Training-Community.pdf

Idahosa, G.E. and Mkhize, Z. (2021) 'Intersectional experiences of Black South African female doctoral students in STEM: participation, success and retention', *Agenda*, 35(2): 110–22.

Joubert, M. (2017) 'White men's voices still dominate public science: here's how to change this', *The Conversation*, Available from: https://theconversation.com/white-mens-voices-still-dominate-public-science-heres-how-to-change-this-96955

Joubert, M. and Mkansi, S. (2020) 'South Africa: science communication throughout turbulent times', in T. Gascoigne, B. Schiele, J. Leach, M. Riedlinger, B. Lewenstein, L. Massarani et al (eds) *Communicating Science: A Global Perspective*, Canberra: ANU Press, pp 771–800.

Kago, G. and Cissé, M. (2022) 'Using African Indigenous languages in science engagement to increase science trust', *Frontiers in Communication*, 6, Available from: https://doi.org/10.3389/fcomm.2021.759069

Karikari, T.K., Yawson, N.A., and E. Quansah (2016) 'Developing science communication in Africa: undergraduate and graduate students should be trained and actively involved in outreach activity development and implementation', *Journal of Undergraduate Neuroscience Education*, 14(2): E5–E8.

Maharajh, R. (2018) 'Towards the next-generation science and technology white paper for South Africa: innovation for transformative change and inclusive development situational analysis', discussion document of the South African National Advisory Council on Innovation, Available from: http://www.naci.org.za/wp-content/uploads/2010/01/Situational-Analysis.pdf

National Research Foundation: NRF (2020) 'National Research Foundation strategy 2020–2025: vision 2030', Available from: https://www.nrf.ac.za/about-us/plans-reports/

National Research Foundation: NRF (2021) 'National Research Foundation information portal: NRF rated researchers', Available from: https://www.nrf.ac.za/information-portal/

Ndlovu, H., Joubert, M., and Boshoff, N. (2016) 'Public science communication in Africa: views and practices of academics at the National University of Science and Technology in Zimbabwe', *Journal of Science Communication* 15(6): A05.

Ramos, A. and Empinotti, M. (2017). 'Indigenous languages must feature more in science communication', *The Conversation*, 19 December, Available from: https://theconversation.com/indigenous-languages-must-feature-more-in-science-communication-88596

Rasekoala, E. (2019) 'The seeming paradox of the need for a feminist agenda for science communication and the notion of science communication as a "ghetto" of women's over-representation: perspectives, interrogations and nuances from the global south', *JCOM: Journal of Science Communication*, 18(4): C07. doi: https://doi.org/10.22323/2.18040307.

Rasekoala, E. (2020a) 'Challenges and opportunities for transformation', in Ecsite webinar 'Decolonising museums and science communication', Available from: https://www.youtube.com/watch?v=CWxw0x7U5lw

Rasekoala, E. (2020b) 'African Gong's conceptualisation of the Africa Scientifique programme', Available from: https://www.youtube.com/watch?v=jnbkidXam80&t=23s

Rasekoala, E. (2022) 'Responsible science communication in Africa: rethinking drivers of policy, Afrocentricity and public engagement', *JCOM: Journal of Science Communication*, 21(4): C01. doi: https://doi.org/10.22323/2.21040301

Rasekoala, E. and African Gong (2019) 'Public health emergencies: the role of science education and communication in Africa', in G.B. Tangwa, A. Abayomi, S. Ujewe, and N. Munung (eds) *Socio-cultural Dimensions of Emerging Infectious Diseases in Africa*, Cham, Switzerland: Springer International, pp 91–107.

Silva, J. and Bultitude, K. (2009) 'Best practice in communications training for public engagement with science, technology, engineering, and mathematics', *Journal of Science Communication*, 8(2): A03.

South African Department of Science and Technology (1996) 'White paper on science & technology: preparing for the 21st century', Available from: https://www.gov.za/sites/default/files/gcis_document/201409/sciencetechnologywhitepaper.pdf

South African Department of Science and Technology (2015) 'Science engagement strategy', Available from: https://www.saasta.ac.za/saasta_wp/wp-content/uploads/2017/11/Science_Engagement_Strategy-11.pdf

South African Department of Science and Technology (2017) 'Science engagement implementation plan', Available from: https://www.dst.gov.za/images/Science_Engagement_Implementation_Plan_13_Nov_2018__copy_002.pdf

South African Department of Science and Technology (2019) 'White paper on science, technology and innovation', Available from: https://www.dst.gov.za/images/2018/white-pate-on-STI-7_09-FINAL.pdf

South African Department of Science and Technology (2020) 'Science engagement monitoring and evaluation framework', Available from: https://www.dst.gov.za/images/DSI_Science_Engagement_MEFFINALREPRONW3b_003.pdf

South African National Development Plan: NDP (2012) 'The National Development Plan: vision for 2030; our future – make it work', Available from: http://www.gov.za/documents/national-development-plan-vision-2030

Statistics South Africa (2021) 'Statistical release: mid-year population estimates 2021', Available from: http://www.statssa.gov.za/publications/P0302/P03022021.pdf

PART II

Science Communication in the Global South: Leveraging Indigenous Knowledge, Cultural Emancipation, and Epistemic Renaissance for Innovative Transformation

5

Challenges of Epistemic Justice and Diversity in Science Communication in Mexico: Imperatives for Radical Re-positioning towards Transformative Contexts of Social Problem-Solving, Cultural Inclusion, and Trans-disciplinarity

Susana Herrera-Lima and Alba Sofía Gutiérrez-Ramírez

Introduction

This chapter seeks to elaborate the challenges that science communication has faced, and continues to face, in Mexico from the perspective of the legacy of colonialism, specifically the condition that decolonial theorists have called 'coloniality of knowledge' (Castro-Gómez, 2000; Quijano, 2000a). These challenges are situated in three complementary dimensions: first, in epistemic coloniality; secondly, in institutional frameworks; and thirdly, in the realms of scientific knowledge and social epistemology. The central premise is that the coloniality of knowledge has determined what kind of knowledge is considered valid and legitimate and therefore worthy of being transmitted and disseminated in institutionalised contexts in the country in the period since its independence in the 19th century to the present day, and that this has implications for the public communication of science. The outline also draws on the premises of the framework of epistemic injustice to examine the role that science communication plays in reproducing practices passed down from the coloniality of knowledge. The chapter concludes by signposting critical recommendations and transformative pathways for a radical reorientation of science communication in Mexico through foregrounding social issues as the point of departure and then explicitly incorporating them into the dialogue, vision, knowledge, and voices of those actors whose knowledge has been historically and systematically excluded.

Modern European science arrived in Mexico as part of a system of knowledge that was considered valid, legitimate, and superior to other ways of knowing. Science communication emerged as a practice passed down from European nations, one that naturalised the validity of scientific knowledge and dismissed the knowledge developed by the original inhabitants as well as their frames of reference and interpretative resources. The argument here is not whether Mexican scientists and communicators have engaged in these practices deliberately; rather, they have been immersed in a system of thinking that validates this hierarchy of knowledge. The intention in this chapter is not to raise doubts about the relevance of scientific knowledge but to question the processes of epistemic imposition and violence, the lack of dialogue, and the delegitimisation of other forms of knowledge, and to challenge what these processes produce as social reality. In brief, the implications these processes have for the way knowledge is generated and worldviews are fashioned in the national sociohistorical context. These worldviews constitute and configure conditions that put certain social groups at a disadvantage in situations of social conflict. This perspective has been neglected in formal studies of science communication.

As noted by Ana Claudia Nepote and Elaine Reynoso-Haynes (2017), science communication in Mexico has been framed as a university-based science activity. It frequently operates to academic agendas that reinforce and complement the scientific education of the population or simply promote an institution's research and its field of science.

Science communication in Mexico inherited a deficitarian perspective, a legacy of the Public Understanding of Science (PUS) movement that emerged in the United States during the post-war years (Lewenstein, 1992) and in Great Britain in the 1980s (Bauer et al, 2006). Present-day discussions in the field have to do with the sociopolitical role of Public Communication of Science (Sánchez-Mora et al, 2015), where even more contemporary approaches argue its value as a cultural strategy aimed at a notion of development that stems from a colonial standpoint. In this contemporary landscape, efforts to include marginal communities in rural areas frequently involve outreach to such populations, yet little effort has been made to build mutual knowledge, understanding, and cross-cultural dialogue (see Patiño et al, 2013).

First dimension: epistemic coloniality

The first dimension that frames the challenges for science communication in Mexico is historical in nature. It refers to the coloniality of knowledge, and more specifically to epistemic colonialism, which dates back to the period of European colonisation in what would later come to be known as Latin America. It was reconfigured as the Mexican nation emerged in the 19th century and assumed a very particular shape throughout the 20th and 21st centuries.

The concept of coloniality was first proposed by Aníbal Quijano, a pioneering Peruvian sociologist who made a key contribution to the fields of dependency and world-systems theory and conceptualised the foundation for decolonial thought and theory:

> Coloniality is one of the constitutive and specific elements of the global pattern of capitalist power. It is based on the imposition of a racial-ethnic classification of the world population as a cornerstone of such a pattern of power, and it operates on each of the material and subjective planes, spheres, and dimensions of social existence, as well as at the social level. (Quijano, 2000a, p 342)

Quijano's seminal work on coloniality, which was first published in 1992 (see Quijano, 1992), gave rise to the modernity/coloniality research agenda and the consequent advocacy for the decolonisation of knowledge that other Latin American thinkers have since developed further based on his original thesis (see Castro-Gómez, 2000; Escobar, 2003; Grosfoguel, 2004; Mignolo, 2007; Restrepo, 2010).

Quijano's work, which he developed together with Immanuel Wallerstein, describes coloniality as a 'form of thinking that established hierarchies between states and societies of the core and those of the periphery … a mental framework that justifies and legitimizes the inequalities between societies in the modern world system' (Quijano and Wallerstein, 1992, p 584. Quoted in Yopasa, 2011).[1] It is a constituent part of the colonial world-system that emerged out of the colonisation of the Americas, rooted in the notion of race and racial hierarchy. Quijano contends that '[s]ocial relations founded on the category of race produced new historical social identities in America … race and racial identity were established as instruments of basic social classification' (Quijano, 2000b, p 534).[2] Overflowing from the political sphere and spilling into the realm of culture, coloniality gave rise to what has been called 'cultural coloniality', which refers to the control of 'the conquered peoples' expressions, knowledge, meanings, symbols, values, practices, and rituals' (Yopasa, 2011, p 115). This in turn led to 'a coloniality regarding the colonized peoples' production of knowledge, their patterns of meaning-making and their symbolic universe' (Yopasa, 2011, p 115).

It is within the framework of this cultural coloniality that epistemic colonialism can be situated. Epistemic colonialism imposes a Eurocentric view of valid knowledge, leaving in its wake imported, institutionalised forms of knowledge that systematically delegitimise local knowledge, particularly that of highly marginalised and vulnerable Indigenous populations. 'Europe's hegemony over the new model of global power concentrated all forms of the control of subjectivity, culture, and especially knowledge and the production of knowledge, under its hegemony' (Quijano, 2000b, p 540).

In Mexico, colonial rule established a regime of truth based on epistemic Eurocentrism and the normalisation of racial differences: Europeans as superior and Indigenous as inferior. This implicitly validated the destruction of the original populations' ways of knowing, understanding, and relating to nature and led to the dismantling of local technologies, methods, and know-how in different fields. The historian Ezequiel Ezcurra has documented, for example, the transformation of land in the central region of what would become New Spain, where the *chinampas* – or floating gardens set amid the network of canals that constituted the city in pre-colonial times – were wiped out by the drainage of Lake Texcoco along with the accumulated knowledge associated with this farming technique (Ezcurra, 2003, pp 9–10). A similar dynamic has been documented in the Jalisco region: in the 19th and 20th centuries, Lake Chapala's surroundings were transformed as large areas were drained to make room for agricultural and industrial development (Boehm, 2005), erasing the earlier ways in which the original population had interacted with the lake.

This situation brings to mind the concept of epistemic violence originally proposed by Gayatri Spivak in the framework of the analysis of subalternity and criticism of Western intellectuals (Spivak, 2009), which, as Manuel Asensi states in the prologue of the critical edition of the text *Can the Subalterns Speak?*, consists of 'thinking of the Other according to a model that in no way explains or accounts for it' (Spivak, 2009, p 18). This concept has been taken up by various theorists in the field of Latin American studies such as Santiago Castro-Gómez, Genara Pulido, and Moira Pérez (Castro-Gómez, 2000; Pulido, 2009; Pérez, 2019). Genara Pulido formulates it as a process that posits the European coloniser as the subject that possesses the explicative truth while the colonised become the Others who wait to be explained, with neither voice nor power, while their forms of knowledge are annihilated in the process. This violence takes the shape of discourses, symbols, and metaphors and is wielded through the imposition of regimes of knowing, as well as the epistemological repression of the Others in the disparagement and invalidation of their knowledge (Pulido, 2009, p 176).

Modern science came to Mexico in the colonial period as part of a system that was seen as valid, legitimate, universal, and superior to other forms of knowledge. Mexican science developed within the epistemological frameworks of European science and its associated criteria of validity. Moreover, during the 19th and early 20th centuries, a vision of science at the service of progress – as defined in Europe – was imported into the country (see Tenorio-Trillo, 1998). Invasive practices that despoiled the environment were legitimised under the aegis of the symbolic authority of scientific knowledge in processes that profoundly transformed urban and rural landscapes, like the 'green revolution' (Ceccon, 2008).

During the second half of the 20th century, this vision of science would be linked to the notion of development, associated with technological and economic development and industrialisation processes. In this period, Mexico's growing economic dependence on the United States was reinforced by a technological dependence that brought with it epistemic dependence. The Mexican nation's heterogeneous cultural configuration encompasses diverse and complex forms of knowledge that are diluted and unrepresented in institutionalised spaces dedicated to the construction and dissemination of knowledge.

Second dimension: institutional frameworks

The challenges facing science communication in Mexico are also tied up with structural factors that determine the conditions in which it takes place: institutional frameworks define and circumscribe the knowledge that is considered valid and therefore worth reproducing and communicating. To a considerable extent, these institutional factors and frameworks end up reinforcing unequal conditions of knowledge production and access and thus de facto practices of exclusion. As Mileidy Yopasa argues, basing her observations on the propositions made by the Latin American critics Quijano and Castro-Gómez, the discovery of the Americas ushered in 'the organization of space and time in their totality, and this construction has as its basic premise the universal character of the European experience, which excludes, denies and defines what is non-European' (Yopasa, 2011, p 114).

This original colonial organisation defined the valid knowledge of the age and left its mark on the institutional frameworks of education in the 19th century, where official programmes were formulated to define and delimit the contents of the knowledge deemed necessary and valid. Local forms of knowledge were ignored and invalidated, both those dating back to before the colonial period and those that endured, developed, and evolved in later periods.

The institutionalisation of scientific knowledge in all its forms played a fundamental role in the configuration of independent Mexico as a modern nation. The coloniality of knowledge also instituted racism and its naturalisation through public discourse, which often resorted to the validation of racial inequality between Europeans and non-Europeans – Indigenous, especially – with references to scientific knowledge. The validation of practices that amounted to social and environmental plunder was upheld by the discourse of progress based on imported scientific and technological development in areas such as agriculture and industry. The 20th century witnessed a gradual normalisation of the erasure of knowledge that was not validated according to the frameworks that define what is scientific, based on official educational programmes at all levels. All things Indigenous were relegated to a past that was exalted but at the same time seen as eclipsed, surpassed, and thus insidiously reduced to folklore in a process of epistemic asphyxiation.

These conditions have barely been examined in the study and analysis of the communication of science in Mexico. Science communication in the country has a rich and longstanding tradition dating back to the 18th century. The following century saw the emergence of museums and scientific publications and societies. The Spanish colonisers gathered information from the collections of flora and fauna and botanical gardens of the societies that already lived in the territory before their arrival; they ordered and classified this information according to the conventions of European science and exhibited the collections in what would be the first Museum of Natural History in Mexico, inaugurated in 1790 (Rico Mansard, 2007, quoted by Reynoso et al, 2020). During the second half of the 19th century, the most important venues for the communication of science and technology in the world were the European World Fairs (Herrera-Lima, 2018a). President Porfirio Díaz's government intentionally proposed the symbolic representation and construction of the Mexican nation through science at these fairs, with the team that organised the Mexican exhibit at the fair called 'the magicians of progress'. Their exhibitions represented a modern nation with an Indigenous past and significant scientific production aligned with the science developed, recognised, and disseminated in Europe. The country already had several scientific societies that sought to correspond with those in Europe (Tenorio Trillo, 1998), thus cementing the legacy of the coloniality of knowledge.

The formal institutionalisation of science communication in Mexico got underway in the 1960s with the development of specialised publications, museums, and programmes, primarily in Mexico City (Reynoso-Haynes et al, 2020). Scientific dissemination in Mexico originally conformed to deficit models, implicitly and unapologetically assuming the framework of thinking and symbolic organisation of the coloniality of knowledge and thus unwittingly imposing the epistemic exclusion of potential audiences in marginalised contexts, both in cities and rural areas.

The anthropology museums of the 1960s presented themselves as the prime opportunity to denaturalise the perceptions and notions associated with the coloniality of knowledge. The National Museum of Anthropology, however, promoted a narrative associated with the construction of the modern nation-state, with a representation of Mexicanness that highlighted the Indigenous Other, once again exalting this period but avoiding the discursive racism put forward in modern anthropological science in the 19th century, as well as the practices of Indigenous slavery in colonial and postcolonial times. The knowledge that circulated before the Spanish conquest was suspended and sacralised in the museum's rarefied space.

This dichotomy has constituted one of the greatest dilemmas and challenges for science communication in the country: on one hand, science communicators undertake a wide range of projects and activities that contribute to the equally wide range of forms of scientific culture in the Mexican population; on the other hand, however, the legacy of the coloniality of knowledge in institutional contexts has left its mark on this scientific culture and contributes to the widening of gaps, epistemic exclusion, and the imposition of frameworks of understanding that are alien to the context of social groups living in isolation and marginalisation without the necessary construction of a common scaffolding of interpretative resources.

Therefore, science communicators develop and work within a framework of institutionalised processes of naturalisation of inequality, of the ongoing construction of the inferior Other (the 'Mexican indio'), and the technologies of subjectification associated with these processes; this in turn gives rise to symbolic battles in which knowledge that lacks the legitimisation of the institutional frameworks is relegated and left in a disadvantaged position that hampers dialogue or a respectful exchange of views. The communicator's work comes up against the chronic dilemma between the concept of 'neutral' and 'non-ideological' science and a diverse and unequal sociocultural context rich in sedimented but historically unrecognised community knowledge.

Third dimension: scientific knowledge and social epistemology

Science communication in Mexico faces another epistemic challenge, one that concerns the scientific knowledge that it foregrounds, and the practices by which it does so. This challenge calls for looking at science communication from the perspective of social epistemology, which considers the ways a knower is positioned within the social world. As Miriam Solomon (1994) explains, this involves a rupture with earlier practices: starting to think about the construction of knowledge as a collective enterprise as opposed to the classic paradigm that Descartes introduced, 'I think therefore I am', which posits individual thinking as the basis of knowledge.

Social epistemology considers the interpersonal and social dimensions of knowledge and the ways in which a knower agent might be embedded in the social world (by race, class, ethnicity, and so on). In the words of Alvin Goldman, who championed social epistemology in the late 1980s: 'In what respects is social epistemology social? Firstly, it focuses on social paths or routes to knowledge. … Secondly social epistemology does not restrict itself to believers taken singly' (Goldman, 1999, pp 4–5). This is of great relevance for this interrogation of the Mexican scenario since science communication is, by definition, a collective practice and because, as Goldman later pointed out, 'social epistemology is theoretically significant because of the central role of society in the knowledge-forming process. It also has practical importance because of its possible role in the redesign of information-related social institutions' (2006, np).

The rationale of social epistemology thus leads to the role of science communicators being reframed not as intermediaries but as mediators in culture. Drawing on the proposals of the Latin American thinker Jesús Martín-Barbero (1990), this role involves making explicit reference to the relationship between cultural difference and social inequality, as well as to the impossibility of thinking about them separately in Mexican society. It implies recognising that

> the meaning of a practice is limited not only by the density or complexity of the text, but by the *situation in which it is read*, and the degree to which it is tied up with other social factors that are not strictly cultural … it calls for assuming that the specificity [of the information or the cultural work] is not made up only of formal differences but also of *references to life worlds and to ways of using*. (Martín-Barbero, 1990, p 14)

On this basis, the public communication of science and its practitioners ought to consider and problematise scientific knowledge as a communication object that concerns knower agents embedded in a social and cultural milieu. One valuable key for this is what Miranda Fricker (2007) calls epistemic injustice.

According to Fricker, an epistemic injustice 'wrongs someone in their capacity as a subject of knowledge, and thus, in a capacity essential to human value' (2007, p 5). She identifies two forms of epistemic injustice: testimonial injustice and hermeneutical injustice. These are the ethical dimensions of two practices in which science communication, as a knowledge-producing practice, operates: gaining knowledge by account of being told, and making sense of our social experiences.

Regarding testimonial injustice, it is relevant to consider the credibility afforded to a person's discourse, and the prejudices deriving from a shared imaginative conception of someone's identity, in social terms. This might undermine the value of their testimony and experience, inflicting a harm 'specifically in her capacity as a knower' (Fricker, 2007, p 20) and, at the same time, giving an unfair advantage to the privileged, who are not subject to such prejudices in communicating their knowledge.

Prejudices operate in science communication due to the epistemic colonialism highlighted earlier, which leads to the disparagement and dismissal of Indigenous communities' knowledge and a disparity between interpretative forms. Mediation can either aggravate or counteract this type of injustice. An exemplar of the latter is that of the experience of working on the case presented by academics from ITESO University[3] before the Latin American Water Tribunal (Tribunal Latinoamericano del Agua [TLA]) regarding the human rights violations in the communities of Mezcala and San Pedro Itzicán, in Jalisco, Mexico.

In this case, a multidisciplinary group was formed by affected members of the Coca community (kidney disease patients) and academics from the Law and Socio-political Department, Sociocultural Studies Department, and Mathematics and Physics Department from ITESO with the goal of enabling the voices and testimonies of the communities, and for the socio and technical elements to take the form of evidence to file a lawsuit on the damage done to the communities' health and well-being. As a result, the TLA delivered a verdict on 26 October 2018 in favour of the plaintiffs, the Coca people affected in the Chapala lakeside. The TLA acknowledged all petitions and ruled on the human rights violations and socio-environmental damage of Lake Chapala and exhorted the Mexican state to fulfil its obligations with regard to the human right to water and sanitation, health, the provision of a healthy environment, and Indigenous peoples' rights (Verduzco-Espinoza, 2019).

Hermeneutical injustice is directly related to interpretative resources, meaning the ideas and concepts that enable people to understand their own social experience. In this type of injustice, the knower 'is wronged in their capacity as a subject of social understanding' (Fricker, 2007, p 7). This is one structural impact of testimonial injustice, affecting what can and cannot be included in the body of collective knowledge and leading to the underrepresentation of the experiences of marginalised groups and individuals, which in turn undermines their ability to make sense of their experiences. An illustrative example can be seen in science museums where these injustices operate, as Marcos Gómez (2018) points out in his analysis of Mayan cultures turned into spectacles in the hegemonic discourse of the museums of the Mayan Riviera in the Mexican state of Quintana Roo, specifically the Maya Museum of Cancún[4] and the Museum of Mayan Culture in Chetumal.[5] By way of contrast, the same author shows how the discourses in community museums constitute a political act where the communities put out a counter-discourse by constructing their own representation of their relationship with nature and their past.

Conclusion: Recommendations for transformation and the advancement of inclusive science communication in Mexico

The expositions and interrogations of the implications of the coloniality of knowledge and epistemic injustices in science communication in Mexico, elaborated in this chapter, contribute to and buttress a series of critical works undertaken in multiple territories of the Global South.

To effect transformative and sustained change in the Mexican scenario, we propose the need for a profound shift in emphasis for science communication, towards the communication of comprehensive knowledge and in favour of cultural inclusion and trans-disciplinarity. This involves a radical shift of the current paradigm, putting social processes in the foreground, where the actors involved identify elements and relations that can be explained and understood by connecting institutionalised scientific knowledge with their own knowledge, constructed from their experience, their history, and their relationship with their surroundings. The aim is to shift the objectives of the communication from an interest in engaging people with scientific knowledge for its own sake to an interest in gaining insight and understanding into social problems for the explicit purpose of contributing to social transformation. 'The objective is not so much to engage the audience with knowledge as to engage scientists and communicators with social causes and problems' (Herrera-Lima, 2018b, p 15).

In Latin America and the Caribbean, authors such as Mónica Lozano (2012) have highlighted the prevalence of linear conceptions of scientific and technological development that have failed to foster social change and development simply by promoting economic productivity and innovation processes. In her words: 'We need to find new paths that will help us produce knowledge that is relevant to society and addresses the issues that communities and regions are grappling with, by recognizing their capacities and knowledge and protecting regions' resources and livelihoods' (Lozano, 2012, p 8).

These and other critiques have given rise to recommendations of good practices for overcoming these limitations. In Latin America, the post-war period in Colombia in the 1990s saw the emergence of the 'social appropriation of knowledge', a perspective for intervening in the relations between scientific knowledge and citizens in a context of advocacy for the democratisation of knowledge, with a strong focus on meeting citizens' needs and addressing social problems. This paradigm has given rise to orientations such as the call for scientific research to work on problems that emerge from social contexts; the recognition of sources of knowledge other than the scientific method; the consideration of systems of values, beliefs, and forms of organisation of the communities where the interventions are occurring;

and the inclusion of models of public participation in science throughout the research process (Lozano et al, 2021).

Beyond Latin America and featuring contributions from other parts of the world, Elizabeth Rasekoala and Lindy Orthia (2020) recommend thinking about science communication history in broader geo-temporal terms in order to combat racism in science communication. Their work has produced two concrete guidelines for cultural inclusion that grew out of a case study of the Indigenous Australian Yorta Yorta Nation. They propose to 'disentangle elements of present-day science communication' and to 'recognize multiple stories'. They argue that science communication ought to be considered as a set of practices with various points of origin that have something in common, and it is through tracing those histories and understanding their distinct elements, their intersections, and differentiations from a genealogical and geographical point of view that they can be stripped of an essentialist notion of 'science communication' based on dominant present-day trends. They take the big picture history out of the single 'global' chronology to avoid the risk of falling into a Eurocentric narrative.

Along these same lines, in their most recent publication, 'From the margins to the mainstream: deconstructing science communication as a White, Western paradigm', Summer Finlay and her collaborators make a categorical denunciation: 'Despite what might seem like a disparate field of actors, it [the science communication 'mainstream'] is a global hegemony in which unexamined knowledge practices are normalized and perpetuated by networks of privileged individuals and well-funded institutions, supported by dominant white, Western cultures. Such power relations must be questioned' (Finlay et al, 2021, p 6).

In the Mexican scenario, new voices and methods in the communication of socio-environmental problems have begun to emerge. At the previously mentioned hearing of the LTA in 2018, struggles over water were aired in public in a space for alternative justice, through exercises of communication that integrated knowledge from different spaces, voices, and sources (Herrera-Lima et al, 2019). The #Yoprefieroellago (I Prefer the Lake) movement in 2018 invoked diverse forms of knowledge regarding the Lake Texcoco ecosystem – degraded by centuries of devastation, pollution, and encroachment – in the struggle for its defence and recovery and proposed a comprehensive restoration of the territory that drew on ancestral knowledge of the region in dialogue with institutionalised knowledge from multiple disciplines. The role of the local population, civil society organisations, and academics and communicators was fundamental in achieving a project that will hopefully lead to the lake's full recovery under the leadership of the current movement #Manosalacuenca.

In addition to these practical orientations that advocate for the radical re-positioning of science communication with multidisciplinary and contextualised mediation in its practices, we also call on the field and on practitioners to be reflexive. The three-dimensional challenge for inclusive science communication in Mexico that has been elaborated in this chapter has developed with respect to the legacy and structures of colonialism and epistemic injustices and prompts the field and its practitioners to engage in a number of lines of critical reflection: we need to ask ourselves what role is played by the identity and credibility of those taking part in science communication processes; we need to take a careful look at the forms of prejudice that operate in them; we need to open up spaces for other voices and testimonies; we need to support the credibility of participants affected by prejudice; and we need to recognise social identities, not just individual identities.

> Ever since I was born I knew about the lake.
> It's my parent's legacy. It's the life we have, the one we want to leave for our grandchildren.
> Anáhuac means to live in the midst of water.
> Fish, frogs and *axolotls*. Ducks, *chichicuilotes* and swallows. Wild greens – *quelites, quintoniles, verdolagas* – for the taking.
> The peoples knew how to live with the lake, collecting *ahuautle* eggs and *tequesquite* salt.
> That lake is called freedom.
>
> (Fragment from *Carta del Lago*, #YoPrefieroElLago)

Notes

1. All citations originally in Spanish have been translated by the authors.
2. Anibal Quijano himself mentions that some of his texts go into the concept of coloniality of power more deeply (Quijano, 1991; 1993; 1994; Quijano and Wallerstein, 1992).
3. Instituto Tecnológico y de Estudios Superiores de Occidente (Western Institute of Technology and Higher Studies). Located in Guadalajara, Jalisco, México.
4. Museo Maya de Cancún. See: https://lugares.inah.gob.mx/es/inicio/expertos/469-museo-maya-de-canc%C3%BAn.html
5. Museo de la Cultura Maya de Chetumal. See: http://sic.gob.mx/ficha.php?table=museo&table_id=106

References

Bauer, M., Allum, N., and Miller, S. (2006) 'What can we learn from 25-years of PUS research? Liberating and widening the agenda', *Public Understanding of Science*, 16(1): 79–95. doi: 10.1177/0963662506071287

Boehm-Schoendube, B. (2005) 'Agua, tecnología y sociedad en la cuenca Lerma-Chapala: una historia regional global', *Nueva antropología*, 19(64): 99–130, Available from: http://www.scielo.org.mx/scielo.php?script=sci_arttext&pid=S0185-06362005000100006&lng=es&tlng=es

Castro-Gómez, S. (2000) 'Ciencias sociales, violencia epistémica y el problema de la "invención del otro"', *La colonialidad del saber: eurocentrismo y ciencias sociales. Perspectivas latinoamericanas*, Buenos Aires: CLACSO, Available from: http://bibliotecavirtual.clacso.org.ar/clacso/sur-sur/20100708045330/8_castro.pdf

Ceccon, E. (2008) 'La revolución verde tragedia en dos actos', *Ciencias*, 1(91): 21–9.

Escobar, A. (2003) 'Mundos y conocimientos de otro modo: el programa de investigación de modernidad/colonialidad Latinoamericano', *Tabula Rasa*, 1: 51–86, Available from: http://www.redalyc.org/articulo.oa?id=39600104

Ezcurra, E. (2003) *De las chinampas a la megalópolis: el medio ambiente en la cuenca de México*, México: Fondo de Cultura Económica.

Finlay, S.M., Raman, S., Rasekoala, E., Mignan, V., Dawson, E., Neeley, L. et al (2021) 'From the margins to the mainstream: deconstructing science communication as a White, Western paradigm', *JCOM: Journal of Science Communication*, 20(1): C02. doi: https://doi.org/10.22323/2.20010302

Fricker, M. (2007) *Epistemic Injustice: Power and the Ethics of Knowing*, Oxford: Oxford University Press.

Goldman, A.I. (1999) *Knowledge in a Social World*, Oxford: Oxford University Press. doi: https://doi.org/10.1093/0198238207.001.0001

Goldman, A. (2006) 'Social epistemology', Stanford Encyclopedia of Philosophy, Available from: https://plato.stanford.edu/archives/sum2010/entries/epistemology-social/

Gómez-Cervantes, M.V. (2018) 'El museo como dispositivo para presentar el discurso de la espectacularización de la naturaleza', in S. Herrera-Lima and C.E. Orozco (eds) *Comunicar ciencia en México: Prácticas y escenarios*, pp 233–61.

Grosfoguel, R. (2004) 'The Implications of Subaltern Epistemologies for Global Capitalism: Transmodernity, Border Thinking and Global Coloniality', in Wi.I. Robinson and P. Applebaum (eds) *Critical Globalization Studies*, London: Routledge, pp 283–92.

Herrera-Lima, S. (2018a) *Del progreso a la armonía: naturaleza, sociedad y discurso en las Exposiciones Universales (1893–2010)*, Tlaquepaque: ITESO.

Herrera-Lima, S. (2018b) 'Voces, narrativas y formas emergentes en comunicación de la ciencia y problemas socioambientales', *JCOM: América Latina* , 1(1): A07. doi: https://doi.org/10.22323/3.01010207

Herrera-Lima, S. (ed) (2019) 'Luchas por el agua y justicia alternativa', *Clavigero: Comunidades de saberes*, no 12, Tlaquepaque: ITESO.

Lewenstein, B. (1992) 'The meaning of public understanding of science in the United States after World War II', *Public Understanding of Science*, 1: 45–68. doi: https://doi.org/10.1088%2F0963-6625%2F1%2F1%2F009

Lozano, M. (2012) 'Tecnología e innovación en América Latina y los desafíos para la democratización de la ciencia', *Trilogía Ciencia Tecnología Sociedad*, 4(6): 7–9. doi: https://doi.org/10.22430/21457778.29

Lozano, M., Mendoza, M., Montaña, D., and Sandoval, R. (2021) 'Apropiación social del conocimiento, investigación participativa y construcción', *Revista Boletín Redipe*, 10: 80–8. doi: https://doi.org/10.36260/rbr.v10i3.1219

Martín-Barbero, J. (1990). 'Comunicación, campo cultural y proyecto mediador', *Diálogos de la Comunicación*, 26: 7–15.

Mignolo, W. (2007) 'Introduction: coloniality of power and de-colonial thinking', *Cultural Studies*, 21(2–3): 155–67. doi: 10.1080/09502380601162498

Nepote, A.C. and Reynoso-Haynes, E. (2017) 'Science communication practices at the National Autonomous University of Mexico', *JCOM: Journal of Science Communication*, 16(5): C05. doi: https://doi.org/10.22323/2.16050305

Patiño, B.L., Padilla, G., Saldívar, C., González, E., and Mata, V. (2013) 'Análisis de la Estrategia Nacional de 2012 de Divulgación de la Ciencia, la Tecnología y la Innovación del CONACYT', *Actas del 'XIX Congreso Nacional de Divulgación de la Ciencia y la Técnica' y 'XIII Reunión de la Red POP*, Zacatecas, México: np, May, pp 20–4.

Pérez, M. (2019) 'Violencia epistémica: reflexiones entre lo invisible y lo ignorable', *El lugar sin límites: Revista de Estudios y Políticas de Género*, 1: 91–8.

Pulido Tirado, G. (2009) 'Violencia epistémica y descolonización del conocimiento', *Sociocriticism*, 24: 173–201.

Quijano, A. (1991) 'Colonialidad y modernidad/racionalidad', *Perú Indígena*, 13(29): 11–29.

Quijano, A. (1992) 'Colonialidad y modernidad/racionalidad', *Perú Indígena*, 13(29): 11–20.

Quijano, A. (1993) 'América Latina en la Economía Mundial', Problemas del desarrollo, 24(95).

Quijano, A. (1994) 'Colonialité du Pouvoir et Democratie en Amerique Latine', Future anterieur: Amerique Latine, democratie et exclusion. Paris, France: L´ Harmattan, pp 51–87.

Quijano, A. (2000a) 'Colonialidad del poder y clasificación social', *Journal of World-Systems Research*, 6(2) (Summer/Fall): 342–86. Special Issue: Festchrift for Immanuel Wallerstein – Part I, Available from: https://jwsr.pitt.edu/ojs/jwsr/article/view/228

Quijano, A. (2000b) 'Coloniality of power, Eurocentrism, and Latin America', *Nepantla: Views from South*, 1(3): 533–80, Available from: https://muse.jhu.edu/article/23906#:~:text=580.%20Project%20MUSE-,muse.jhu.edu/article/23906,-

Quijano, A. and Wallerstein, I. (1992) 'La americanidad como concepto, o América en el moderno sistema mundial', *Revista Internacional de Ciencias Sociales*, 44(4): 583–93.

Rasekoala, E. and Orthia, L. (2020) 'Anti-racist science communication starts with recognising its globally diverse historical footprint', Impact of Social Sciences, Available from: https://blogs.lse.ac.uk/impactofsocialsciences/2020/07/01/anti-racist-science-communication-starts-with-recognising-its-globally-diverse-historical-footprint/

Restrepo, E. and Rojas, A. (2010) *Inflexión decolonial: fuentes, conceptos y cuestionamientos*, Popayán, Colombia: Editorial Universidad del Cauca.

Reynoso-Haynes, E., Herrera-Lima, S., Nepote, A., and Patiño, L. (2020) 'From simple and centralised to expansion, diversity and complexity', in T. Gascoigne, B. Schiele, J. Leach, M. Riedlinger, B. Lewenstein, L. Massarani et al (eds) *Communicating Science: A Global Perspective*, Canberra: ANU Press, pp 567–596.

Rico-Mansard, L. (2007) 'La historia natural tras las vitrinas', in L.F. Rico, M.C. Sánchez, J. Tagüeña, J. Tonda (eds) *Museología de la ciencia: 15 años de experiencia*, México: Dirección General de Divulgación de la Ciencia, Universidad Nacional Autónoma de México, pp 37–66.

Sánchez-Mora, C., Reynoso-Haynes, E., Mora, A.M.S., and Parga, J.T. (2015) 'Public communication of science in Mexico: past, present and future of a profession', *Public Understanding of Science*, 24(1): 38–52. doi: 10.1177/0963662514527204

Solomon, M. (1994) 'Social empiricism', *Noûs*, 28(3): 325–43. doi: https://doi.org/10.2307/2216062

Spivak, G. (2009) *¿Pueden hablar los subalternos? Traducción y edición crítica de Manuel Asensi Pérez*, Barcelona: Museu D'Art Contemporani de Barcelona.

Tenorio-Trillo, M. (1998) *Artilugio de la nación moderna: México en las exposiciones universales, 1880–1930*, México: Fondo de Cultura Económica.

Verduzco-Espinoza, A. (2019) 'La configuración de la desigualdad de conocimiento sociotécnico en los procesos de litigación de conflictos socioambientales: los casos Chapala y El Zapotillo', PhD dissertation, Instituto Tecnológico y de Estudios Superiores de Occidente, Available from: http://rei.iteso.mx/handle/11117/6081

Yopasa, M. (2011) 'Geopolítica del conocimiento en América Latina: la construcción de espacios históricos otros', *Revista Austral de Ciencias Sociales*, 21: 111–36. doi: https://doi.org/10.4206/rev.austral.cienc.soc.2011.n21-06

6

Past, Present, and Future: Perspectives on the Development of an Indigenous Science Communication Agenda in Nigeria

Temilade Sesan and Ayodele Ibiyemi

Introduction

This chapter examines the state of science communication in Nigeria and offers perspectives on the development of an inclusive, homegrown science communication agenda. Critical points in the development of the science, technology, and innovation (STI) agenda in Nigeria and, indeed, Africa, correspond to the occurrence of major shifts in history. In pre-colonial times, each society had its own ways of disseminating its ideas of science, which ensured that scientific practices were handed down from one generation to another. In the colonial period, science education was like every other form of education that was nascent at the time: it had as one of its goals the creation of a Western-educated class of people who could function within the colonial system. The wave of pan-Africanism that swept through the continent in the period leading to independence and the period immediately after it inspired a new wave of Afrocentric policies that seemed set to put science and, by extension, science communication, on a transformative path unencumbered by the age-long tendency to refract African scientific endeavour through European lenses.

Over time, however, the challenges of delivering on these laudable visions of an African-led science education, knowledge production, and science communication agenda in former colonies such as Nigeria have become glaring in the face of the legacy and sustained hegemony of dominant Eurocentric systems of knowledge production, validation, and communication (see Alatas, 2006; Asante, 2011; Aman, 2018; R'boul, 2021). This chapter provides an overview of past and present developments on the Nigerian science communication scene against the background of these longstanding challenges. Further, the chapter highlights opportunities for deepening inclusion and relevance towards the goal of building a transformative national science communication agenda for the future.

Rhetoric versus reality: science communication in Nigeria's STI policy and National Innovation System

The first thing a casual observer might notice about the popular narrative around science communication in Nigeria is that it is framed as being indispensable to achieving the goal of national development. By many local accounts, the advancement of science and the public's embrace of its achievements are all-important for unlocking the potential for broad-based innovation and wealth that lie dormant and largely untapped – a far cry from the unidirectional focus on oil that the country has exhibited since 1960. This expectation of science is not peculiar to Nigeria, and it is not new – it dates back to Kwame Nkrumah of Ghana, who laid out a pan-African vision for a science-led future for the continent at a summit of the Organisation of African Unity in 1963:

> We shall accumulate machinery and establish steel works, iron foundries and factories; we shall link the various States of our continent with communications; we shall astound the world with our hydroelectric power; we shall drain marshes and swamps, clear infested areas, feed the under-nourished, and rid our people of parasites and disease. It is within the possibility of science and technology to make even the Sahara bloom into a vast field with verdant vegetation for agricultural and industrial developments. (African Union, 1963: 48)

Political leaders and policy makers in Nigeria agree with this vision: according to Siyanbola et al (2016), there is a consensus among all stakeholders that science and technology are integral to the task of nation-building, particularly in the economic sense. This led the Federal Ministry of Science and Technology (FMST) to publish an STI policy in 2012, the first such policy in the country to be informed by robust evidence (Siyanbola et al, 2016). In keeping with widely held neoliberal assumptions, the STI policy emphasises the role of innovation in linking scientific discovery to market mechanisms in order to drive the kind of economic growth and development envisaged for the country.

Notwithstanding the normative role ascribed to science and innovation, however, Siyanbola et al (2016) note the absence of communication between Nigerian policy makers and legislators on the one hand and scientific actors that generate evidence and produce policy-relevant outputs on the other. This is despite the existence of several public institutions – key among them the Sheda Science and Technology Complex (SHESTCO),[1] the National Centre for Technology Management (NACETEM),[2] the National Office for Technology Acquisition and Promotion (NOTAP),[3] and indeed, the FMST – that are well positioned to facilitate science policy engagement. Consequently, there is a huge gap between the promise of science to deliver development gains to Nigerian society and the current realities of its citizens (Africasti, 2021; Sesan and Siyanbola, 2021).

The 2012 STI policy, in addition to its emphasis on science-led innovation and development, recognises the need for communicating advances in science and technology to the public through 'promotion' activities. In articulating the strategies for achieving this vision of science promotion, the policy covers many bases, including the diversification of communication platforms, the building of local capacity for doing science that addresses local needs, the inclusion of women and youth in communication activities, and the use of local languages for these activities. However, as Falade et al (2020) note, the Nigerian government is far from realising many of the objectives and strategies laid out in the policy.

The Nigerian government's poor record with respect to science communication is not necessarily an anomaly. Gascoigne and Schiele (2020) point out that, even though most governments have been interested in the premise of harnessing science for development since at least the end of the Second World War, the record on those governments actually following through on their professed belief in the potential of science communication is mixed. Still, successful examples of governments taking the initiative outside of Africa – as in the case of South Korea – demonstrate how nations can indeed turn their fortunes around within a relatively short period when governments are at the forefront of efforts to seed technological as well as cultural change through science communication (Gascoigne and Schiele, 2020).

Government ownership of the science communication agenda is arguably more relevant in a country like Nigeria where science has to contend with traditional values, religious beliefs, myths, and superstition to a greater degree than obtains elsewhere (Falade, 2015). In the absence of coordinated government action and leadership in science communication, citizens have not come to see science as an integral part of the culture – as shaping collective notions of 'common sense', as Falade (2015) aptly describes it. Generally, however, national policies around science communication, and the degree to which they are implemented, are themselves influenced by the wider culture (Gascoigne and Schiele, 2020), creating a catch-22 situation in the case of Nigeria.

Far from being embraced as an element of national culture, there is a history in Nigeria of science being associated with the West – so that the public is quick to turn seemingly isolated episodes of scientific uncertainty into broad-ranging conspiracy theories that cast local populations as unwitting victims of Western scientific manipulation (Falade, 2019). These tensions between science and citizens have been most visible in public health. The public's deep-rooted distrust of science is partly the result of multi-generational dissonance in which non-European populations have become othered in the production and communication of science (even though Africa in particular played a pioneering role in scientific innovation – see Maina, 2019). Thus, science has come to be synonymous with the West rather than being understood as the diverse global enterprise that it is (Gunaratne, 2009; Okere et al, 2011). There is also the issue of inadequate representation, which means that many of those involved in science locally – either as scientists or as communicators – are not visible in the public sphere (Maina, 2018).

It is apparent at this point that a lot more needs to be done – by the government as well as other stakeholders – to develop a homegrown scientific agenda through coordinated, consistent, and effective science communication policies and practices. Nonetheless, the dominant discourse around the need for public understanding of science in Nigeria is tinged with more than a bit of scientific hubris. Underlying this discourse is the assumption that the superiority of a scientific worldview is self-evident and that it is up to the political and scientific elite to 'educate' the public on the benefits of aligning with that worldview (Olorundare, 1988; Tsanni, 2019). In reality, science communication in Nigeria needs to be framed as a two-way exchange between scientific actors and the public if it is to be effective and inclusive. First, though, this chapter traces the broad arc of science communication as it has unravelled in Nigeria in postcolonial times, most notably in the domains of agriculture and health.

The broad arc of science communication in Nigeria: from agricultural extension to COVID-19 containment

The earliest recorded instances of science communication in Nigeria occurred in the field of agriculture, in the form of extension services provided by government extension agents and research institutions to farmers in rural areas of the country (Falade et al, 2020). The government's early interest in science-led agricultural development can perhaps be explained by the fact that this was the era before the oil boom, when agriculture still played a major role in the country's domestic and global competitiveness (Falade et al, 2020). The period featured successive programmes of science communication and public engagement aimed at introducing new and improved agricultural practices to farming communities and promoting their subsequent uptake (Falade et al, 2020). Those activities continue to date, albeit in a somewhat muted form, and the range of communication channels has broadened from traditional sources such as radio and print media to include newer platforms such as mobile phones and the internet. Nonetheless, Mgbenka et al (2013) maintain that the most appropriate and effective channels remain those that enable personal exchanges between farmers and scientists, underscoring the significance of mutual communication highlighted earlier. Related to this is the intractability of the use of English as the 'official' language of both science and communication in the country: Lawal (2015) argues that many of the bottlenecks in agricultural development could be resolved if local languages were mainstreamed into science communication activities.

The other sector that has a long history of deploying science in public engagement in Nigeria is that of health. Over the years, government and civil society actors have launched popular multimedia campaigns – some of them now fairly routine – communicating strategies for preventing and treating familiar but deadly diseases such as malaria and polio. Visual evidence of these campaigns abound – in the ubiquitous posters displayed prominently in primary health centres in rural areas, and on giant billboards depicting happy faces attesting to the efficacy of the latest World Health Organization – recommended insecticide spray or antimalarial treatment. Social media platforms are also becoming increasingly deployed in health communication by private and public actors. In a prominent example, the Nigeria Centre for Disease Control (NCDC) keeps a record of infections and fatalities from the ongoing COVID-19 pandemic, which has been converted into an easily recognisable graphic that is widely circulated via Twitter and WhatsApp. This visual report, published daily since March 2020, has become a source of clear and consistent information for the public in the midst of a continuously shifting landscape.

However, perhaps the most instructive instances of science communication in the sector are the ones that developed in response to a backlash from citizens doubtful of the intentions of state and market actors (often with links to Western institutions) implementing public health campaigns – thus exemplifying the distrust towards science noted earlier in the chapter. Crucially, this distrust occurred within the broader context of alienation between the majority of citizens and orthodox health institutions (WHO, 2002). High-profile cases featured accusations of bad faith and a hidden agenda of sterilisation in a 1996 meningitis treatment campaign (run by Pfizer, the American pharmaceutical giant, in collaboration with Johns Hopkins University) and a 2003 government-directed polio vaccination drive (see Jegede, 2007; Waisbord et al, 2010; Falade, 2015). Further proof of public resistance to health messaging was seen during the Ebola epidemic that broke out in West Africa in 2014, when health authorities faced an uphill task in trying to persuade citizens to modify their social practices in line with scientific advice, and progress was only made when the advice was channelled through the traditional and social networks that the people value (Finlay et al, 2021).

These instances highlight both the limits and opportunities for science communication in the Nigerian context: while it may be the case that large swathes of the public do not interact with science in prescribed ways, there is an opportunity to reframe the discussion and engage with citizens on terms that they find more accessible and meaningful. In this regard, it is illuminating at this juncture to undertake a historical journey back to the dynamics of pre-colonial knowledge production and dissemination paradigms in Nigerian society. Fafunwa (1974) notes that functionalism was the main guiding principle of knowledge generation back then across much of the African continent. All knowledge produced was for specific functional purposes, and every society had its own praxis that was influenced by its history, geography, and culture. Furthermore, knowledge was both produced and disseminated through various means that were attuned to a community's needs. Knowledge was produced and disseminated by practice, yet it was structured and systematic. Knowledge was also disseminated through pep talks, age group meetings, moonlight storytelling sessions, and rites of passage. Societies usually codified their knowledge in proverbs, myths, and folktales (see, for example, Yankah, 1995), which, in turn, acted as veritable vehicles of dissemination and education among the people (Majasan, 1969).

One possible agenda for the 'Indigenisation' of science communication in Nigeria would involve contextualising these historical socioculturally embedded practices in a way that recognises their continued relevance for many Nigerians today, particularly those living in rural areas (Rasekoala, 2022). Before discussing this further, however, we need to sketch the outlines of the contemporary science communication landscape in the country, highlighting the key actors and activities involved.

Science communication initiatives in Nigeria: actors, activities, and associations

It is clear from the preceding discussion that a science communication paradigm – one that supports the deliberate cultivation of a homegrown science culture among the public – is not embedded in Nigerian policy and practice. Indeed, according to Maina (2019), scientific literacy among the Nigerian population is extremely low, barely reaching 10 per cent. It is also apparent, however, that this does not translate into a total absence of science communication activity; rather, the main challenge is that there has been little coordination and expansion of existing initiatives, resulting in large part from a long-term failure to implement key aspects of the Nigerian STI policy referenced earlier in the chapter.

Using Gascoigne and Schiele's (2020) broad conceptualisation of science communication as involving activities designed to 'inform, engage, persuade, change behaviours and support better decision-making' (p 12), it is possible to identify two main streams of science communication activity in Nigeria. These are, respectively, activities aimed at making science education in general more desirable and accessible, particularly to young people (popularisation of science education); and activities focused on distilling and disseminating often complex scientific information to address specific public interest challenges (popularisation of science advice). The discussion here focuses on the latter.

Science communication activity in Nigeria comprises initiatives – many of them run by individuals – that convert scientific information on specific issues into formats that are easily accessible and digestible by the general public. The typical vehicle employed by communicators in this category is the written word (in print as well as online), locating much of their activity squarely within the field of science journalism. Accordingly, these initiatives tend to be led by trained scientists who branch out into science reporting and communication or trained journalists who devote their careers to covering science news. Increasingly, however, written communication, often in the form of articles and news bulletins, is being complemented by audio and video programming, especially on online platforms.

Prominent science advice outlets include time-tested platforms like the Development Communications Network, which explicitly links science communication to traditional development sectors such as agriculture, health, education, and environment; and Africa STI News, which similarly reports on a broad range of scientific topics and sees a role for itself in 'showcasing African products, research and innovation to the world'.[4]

Several journalistic platforms have joined the fray in recent years, opening the field up to a younger generation of science communicators. These include Science Nigeria, a homegrown online platform that reports on developments in science, technology, and innovation; Science Communication Hub (SciCom Hub), an online community of scientists and journalists united in their purpose of communicating scientific advances to the public; and the African Science Literacy Network (ASLN), which seeks to develop the capacity of scientists and journalists to collaborate on communication and public engagement initiatives. SciCom Hub and ASLN are notable for their links to the Teaching and Research in Natural Sciences for Development (TReND) in Africa initiative,[5] which originated in Cambridge University, UK, and is supported primarily by Western research and funding institutions. This demonstrates the extent to which Western funding and support still figure in the domestic science communication landscape. While these platforms are important for the way they address a crucial gap in the Nigerian context, they still exist on the fringes, indicating the need for more concrete support at the local and national levels.

Perhaps the most institutionalised response to the gap in this area is the science communication curriculum recently developed by the Institute of Strategic and Development Communication at Nasarawa State University (NSUK) with support from SciDev.Net, the global science communication network, and Robert Bosch Stiftung, a leading German foundation that funds development causes in Africa and elsewhere.[6] Whereas mass communication and development communication courses have long been taught in Nigerian universities, the NSUK programme is the first in the country to focus specifically on science communication (Ologunagba, 2019). The programme, which is offered at both the undergraduate and postgraduate levels, targets not only journalism majors but also students enrolled on other arts and humanities courses such as public relations and performing arts (Ologunagba, 2019). The programme therefore goes beyond the traditional focus on journalism, potentially laying the foundation for greater diversity in the practice of science communication in the future.

The foregoing description of the science communication landscape in Nigeria reveals a system that has great potential but has received very little by way of nurturing, especially from within. In the next section, we propose several ways in which local actors can begin to take ownership of the science communication agenda in order to realise outcomes that are both locally relevant and globally competitive.

The way forward: an agenda for Indigenous science communication in Nigeria

Strengthening the teaching and practice of science

It is not possible to talk about the state of science communication in Nigeria without referring to the state of the science education infrastructure that underlies it. From the primary to the tertiary levels, fundamental gaps in the content and delivery of science education inhibit the development of a solid base upon which effective science communication can be built. The state of science education in the country makes science communication less substantive, and vice versa. While some good research has been produced at the higher levels – for example, in the area of traditional medicine – there is still a lot of substandard research being passed off as science, as local scientists themselves point out (Zanna, 2019). The good news is that there are already frameworks in place for building a more robust science ecosystem. For example, SHESTCO, referred to earlier, operates out of a sprawling facility and has its sights trained on every issue of national significance, from local agriculture to climate change and atomic energy. Importantly, it has the backing of the executive and legislative arms of government, at least in theory. However, this scientific institution, like several others, is operating below its potential, with acute staff shortages and low levels of societal engagement. The immediate task is to build technical and outreach capacity in these institutions so that they become truly indispensable in the National Innovation System and are well integrated into society.

Make adequate funding available for science education and communication

A major challenge associated with the problem of poor infrastructure is that of inadequate funding for science education and communication in the country (Maina, 2019). This is an unfortunate state of affairs, as effective science communication has the potential to help attract funding and other support for science education – the more aware citizens are of important scientific advances that affect their lives, the more likely they will be to prioritise science as a personal and political issue (Nigeria Health Watch, 2019). However, as Falade et al (2020) point out, the growth of science communication in the country is hampered by insufficient support from within, so that local actors are better off looking to Western sources to fund their agendas. It is disconcerting to note the significant gap between the funding earmarked for science communication in national budgets, as described earlier, and actual infrastructure on the ground. Yet, it is precisely this investment in science communication at the national level, as demonstrated by the use of allocated funds to build and sustain infrastructure, that is required if the country is to seed a science culture that will germinate and permeate society for generations to come.

Elevate local scientists – both female and male – as role models in society

The outreach work done by the local scientists who volunteer with TReND in Africa illustrates the importance of validating the aspirations of young African students by introducing them to accomplished scientists who look just like them – for them to see, as Maina (2019) put it, that science 'does not have a colour'. This is a component of science communication that needs to be given more weight in the Nigerian context. Several examples exist of how to do this in practice. One model that is close to home is provided by the Next Einstein Forum (NEF), a regional platform that, notwithstanding the Western reference embedded in its name, is dedicated to showcasing the work and personalities of African scientists on the global stage. Through the outreach activities of its network of fellows and ambassadors, NEF aims to spur interest in, and build capacity for, African science, ultimately drawing it from the fringes into the mainstream. A local expression of this kind of programme has the potential to make it even more responsive to the national context. It will be important to pay close attention to the gender balance of role models on such a programme, as participation in science in the country skews overwhelmingly male (Okeke, 2019). A preponderance of female role models in science is required to address this imbalance, which begins to take shape from the early years of education and persists well into the career phase (Ekine, 2013).

Promote the use of local languages in science communication

The dominance of English and other European languages as the medium of science communication globally has had exclusionary effects among local populations, in Nigeria and elsewhere. Ramos and Empinotti (2017) make a case for prioritising quality – the diversity of audiences – over quantity – a focus on pure numbers – in measuring the effectiveness of local science communication initiatives. The opportunity to promote diversity comes with a challenge in Nigeria, where over 500 local languages are spoken across the country (Statista, 2020). Nonetheless, starting with a campaign to integrate the most widely spoken languages – Hausa/Fulani (35.1 per cent), Yoruba (17.4 per cent), Igbo (9.5 per cent), and Pidgin English (3 per cent) – into science communication programmes would constitute a big step forward. This step requires self-belief: a conviction that science belongs in the realm of the local, and that the latter therefore has something valuable to offer the former. Beyond a focus on language, cultural mediation will be necessary to more fully translate between scientific and non-scientific actors and consequently minimise conflict. In terms of policy, the science, technology, engineering, arts, and mathematics (STEAM) initiative of the federal government[7] presents an opportunity to tie linguistics, which is a core component of the arts and humanities, into the teaching and communication of science, technology, engineering, and mathematics concepts at all levels.

Diversify and popularise communication channels

While the diversity of local platforms for science communication has increased in recent times, especially with the emergence of a new generation of scientists and practitioners, there is still a considerable way to go in employing mediums that are available and accessible to broader swathes of the Nigerian population. The country's science communication field can sometimes appear insular, with practitioners engaging on forums that are largely attended by people who are already on board and do not need to be persuaded of the significance of the enterprise. Practitioners need to pay attention to the nature of the outlets that intended audiences, be they publics or policy makers, engage with in the natural course of their lives – and then target those outlets for science communication. Additionally, a lot more can be done to increase the visibility of existing niche platforms. Increased funding to these platforms would help, but so would a heavy dose of creativity. As has been demonstrated by the NCDC COVID-19 information campaign, which began on Twitter but has taken on a life of its own on platforms such as WhatsApp, social media in particular presents an opportunity for communicators to amplify scientific messaging across many different audience types.

Address the hubris implicit in science communication language and practice

There is a pressing need to bridge the power distance between what essentially constitutes a scientific elite in Nigeria and the wider public. There is a tendency for this elite to be dismissive of the concerns and hesitations of the public around scientific issues on the basis that the presumed authority of science ought not to be questioned by those on whose behalf it is being done. Scholars in the field of science and technology studies have long challenged this 'deficit model' of science communication, which casts the public as uncomprehending laypeople who will shed their inhibitions about particular issues once they are provided with accurate scientific information (Reincke et al, 2020). Missing from this characterisation is the recognition that publics have beliefs and value systems that may vie with science for primacy, and that the scientific method is only one of several possible ways of knowing about the world. Rather than assuming a superior position for science a priori, science communicators need to display a willingness to engage with citizens using terms they can relate to. Health emergencies like the Ebola and vaccine misinformation cases cited earlier have demonstrated that the greatest advances in changing public attitudes are made when scientists find common ground with the public and work from there. The science communication field would benefit from institutionalising this respectful approach, not just in health but in all domains of science and technology.

Conclusion

There are several challenges to the vision of developing a homegrown science communication agenda in Nigeria. Some of these challenges have been highlighted in the current chapter, along with some possible ways of mitigating them. A recurring problem is the adoption of communication paradigms that alienate the citizenry and spur tensions between scientific and non-scientific actors. Past and present attempts to bridge the gap between science communication and Nigerian society, though well intentioned, have not been entirely successful, partly because they have been out of touch with the needs and realities of ordinary citizens.

Furthermore, there are communication gaps that make citizens susceptible to misinformation, as key stakeholders and government agencies have failed to step up and take charge of the national science communication narrative. For the most part, local initiatives that are driven by private individuals and corporate bodies lack the support of government at the local and national levels. More often than not, these private initiatives are concentrated in the hands of technical 'experts', scientists, scholars, and science journalists. This is a missed opportunity to mainstream the perspectives and practices of lay publics across the country into science communication so that it goes beyond the usual calls for dissemination of knowledge to a passive citizenry.

In conclusion, there is a dire need – but also significant potential – for scientific and non-scientific actors to jointly develop a truly inclusive and transformative science communication agenda for Nigeria, one that will be better equipped than past and present models to deliver broad-based development to citizens at every level of society.

Notes

[1] SHESTCO (www.shestco.gov.ng) is a national research institution with a mandate to develop homegrown technological solutions to Nigeria's problems and showcase them on the global stage; however, productivity has been hampered by institutional problems including poor management and low levels of funding.

[2] NACETEM (www.nacetem.gov.ng) was established to conduct policy research and provide technical advice to the federal government in the area of science and technology management. NACETEM provided significant input to the 2012 STI policy, including the development of evidence-informed indicators that reflect the realities of the National Innovation System.

[3] NOTAP (www.notap.gov.ng) is the government agency responsible for identifying promising technologies developed by local innovators and supporting them in their efforts to commercialise those technologies.

[4] Africa STI News website: http://africasti.com.ng/

[5] TReND (https://trendinafrica.org/) is an example of a regional initiative that has found local expression in Nigeria, most pertinently through its outreach programme. In the absence of strong government leadership in science communication in the country, non-governmental organisations like TReND have stepped in with activities aimed at helping students and teachers alike to engage with basic science content in interactive and innovative ways, including 'hands-on' workshops, radio shows, science festivals, and role modelling (Zanna, 2019). By its own account, the TReND initiative has succeeded in attracting thousands of young people across Africa to the natural sciences and providing them with the training and tools they need to pursue their interests – all while building their confidence by tapping into the pool of local researchers for the implementation of outreach activities. While initiatives like this do make a difference, it is clear that they need to be scaled up significantly for their impact to be truly inclusive and transformative.

[6] The curriculum was developed under Robert Bosch's Script project, which provides free training and networking opportunities to budding science communicators across Africa. Other partners within the Script network include Makerere University (Uganda), Moi University (Kenya), University of Dar Es Salaam (Tanzania), Federal Radio Corporation Nigeria, and The Conversation Africa.

[7] Departing from its historical focus on STEM education, the federal government is now making an effort to include the arts and humanities in the narrative on science-led development, and policy makers at the highest level now talk of the need to strengthen STEAM education in the country – the 'A' standing for 'arts'.

References

African Union (1963) 'Speeches and statements made at the first Organization of African Unity (O.A.U.) summit', Available from: https://au.int/sites/default/files/speeches/38523-sp-oau_summit_may_1963_speeches.pdf

Africasti (2021) 'Nigerian scientists should use their works to link society with governance', Available from: https://ntm.ng/2021/06/25/scientists-should-use-their-works-to-link-society-with-governance/

Alatas, S.F. (2006) *Alternative Discourses in Asian Social Science*, New Delhi: Sage.

Aman, R. (2018) *Decolonising Intercultural Education*, London: Routledge.

Asante, M.K. (2011) 'De-Westernizing communication: strategies for neutralizing cultural myths', in G. Wang (ed) *De-Westernizing Communication Research: Altering Questions and Changing Frameworks*, London: Routledge, pp 21–7.

Ekine, A. (2013) 'Enhancing girls' participation in science in Nigeria: Improving Learning Opportunities and Outcomes for Girls in Africa', *Brookings Institution*, 41(3): 12–26, Available at: https://www.brookings.edu/wp-content/uploads/2016/07/ekine_girls_education.pdf

Fafunwa B. (1974) *History of Education in Nigeria*, Oxon: Routledge.

Falade, B. (2015) 'Familiarising science: a Western conspiracy and the vaccination revolt in Northern Nigeria', *Papers on Social Representations*, 24(3): 1–24, Available from: http://www.psych.lse.ac.uk/psr/

Falade, B. (2019) 'Religious and traditional belief systems coexist and compete with science for cultural authority in West Africa', *Cultures of Science*, 2(1): 9–22.

Falade, B., Batta, H., and Onifade, D. (2020) 'Nigeria: battling the odds; science communication in an African state', in T. Gascoigne, B. Schiele, J. Leach, M. Riedlinger, B.V. Lewenstein, L. Massarani, et al (eds) *Communicating Science: A Global Perspective*, Canberra: ANU Press, pp 615–40.

Finlay, S.M., Raman, S., Rasekoala, E., Mignan, V., Dawson, E., Neeley, L. et al (2021) 'From the margins to the mainstream: deconstructing science communication as a White, Western paradigm', *Journal of Science Communication*, 20(1): C02. doi: 10.22323/2.20010302

Gascoigne, T. and Schiele, B. (2020) 'Introduction: a global trend, an emerging field, a multiplicity of understandings; science communication in 39 countries', in T. Gascoigne, B. Schiele, J. Leach, M. Riedlinger, B.V. Lewenstein, L. Massarani et al (eds) *Communicating Science: A Global Perspective*, Canberra: ANU Press, pp 1–14.

Gunaratne, S.A. (2009) 'Globalization: a non-Western perspective; the bias of social science/communication oligopoly', *Communication, Culture & Critique*, 2(1): 60–82.

Jegede, A.S. (2007) 'What led to the Nigerian boycott of the polio vaccination campaign?', *PLoS Medicine*, 4(3): e73. doi: 10.1371/journal.pmed.0040073

Lawal, O.A. (2015) 'Indigenous languages as tools for effective communication of science and technology for food production in Nigeria', *Theory and Practice in Language Studies*, 5(3): 463–8. doi: 10.17507/tpls.0503.02

Maina, M. (2018) 'Online science campaign to inspire the next generation of African scientists', *The Biochemist*, 40(6): 38–40. doi: 10.1042/BIO04006038

Maina, M. (2019) 'Inspiring a next generation science in Africa through science communication', Brussels: FENS-Kavli Network of Excellence, Available from: https://fenskavlinetwork.org/inspiring-a-next-generation-science-in-africa-through-science-communication/

Majasan, J. (1969) 'Folklore as an instrument of education among the Yoruba', *Folklore*, 80: 41–59.

Mgbenka, R.N., Agwu, A.E., and Ajani, E.N. (2013) 'Communication platforms existing among researchers, extension workers, and farmers in Eastern Nigeria', *Journal of Agricultural & Food Information*, 14(3): 242–58. doi: 10.1080/10496505.2013.808928

Nigeria Health Watch (2019) 'Telling the whole story: better science communication for national development', 7 November, Available from: https://nigeriahealthwatch.com/telling-the-whole-story-better-science-communication-for-national-development/

Okeke, F.N. (2019) 'Challenges in Nigeria, to attract more girls and women to STEM career', Geneva: United Nations Conference on Trade and Development.

Okere, T., Njoku, C.A., and Devisch, R. (2011) 'All knowledge is first of all local knowledge', in R. Devisch and F. Nyamnjoh (eds) *The Postcolonial Turn*, Bamenda: Langaa and African Studies Centre, pp 275–96.

Ologunagba, C. (2019) 'Nasarawa varsity popularises science communication curriculum in other universities', *NNN*, 28 October, Available from: https://nnn.ng/nasarawa-varsity-popularises-science-communication-curriculum-in-other-universities-don/

Olorundare, S.A. (1988) 'Scientific literacy in Nigeria: the role of science education programmes', *International Journal of Science Education*, 10(2): 151–8. doi: 10.1080/0950069880100203

Ramos, A. and Empinotti, M. (2017) 'Indigenous languages must feature more in science communication', *The Conversation*, 19 December, Available from: https://theconversation.com/indigenous-languages-must-feature-more-in-science-communication-88596

R'boul, H. (2021) 'North/South imbalances in intercultural communication education', *Language and Intercultural Communication*, 21(2): 144–57. doi: 10.1080/14708477.2020.1866593

Rasekoala, E. (2022) 'Responsible science communication in Africa: rethinking drivers of policy, Afrocentricity and public engagement', *JCOM: Journal of Science Communication*, 21(4): C01. doi: https://doi.org/10.22323/2.21040301

Reincke, C.M., Bredenoord, A.L., and van Mil, M.H. (2020) 'From deficit to dialogue in science communication: the dialogue communication model requires additional roles from scientists', *EMBO Reports*, 2(9): e51278. doi: 10.15252/embr.202051278

Sesan, T. and Siyanbola, W. (2021) '"These are the realities": insights from facilitating researcher–policymaker engagement in Nigeria's household energy sector', *Humanities and Social Sciences Communications*, 8(73): 1–11. doi: 10.1057/s41599-021-00754-5

Siyanbola, W., Adeyeye, A., Olaopa, O., and Hassan, O. (2016) 'Science, technology and innovation indicators in policy-making: the Nigerian experience', *Palgrave Communications*, 2(16015). doi: 10.1057/palcomms.2016.15

Statista (2020) 'Primary languages spoken at home in Nigeria as of 2020', Available from: https://www.statista.com/statistics/1268798/main-languages-spoken-at-home-in-nigeria/

Tagoe, H.A. and Tagoe, T.A. (2020) 'Ghana: when individuals refuse to let science communication die', in T. Gascoigne, B. Schiele, J. Leach, M. Riedlinger, B.V. Lewenstein, L. Massarani et al (eds) *Communicating Science: A Global Perspective*, Canberra: ANU Press, pp 351–69.

Tsanni, A. (2019) 'African Science Literacy Network: science communication and journalism workshop holds in Abuja', 12 August, Available from: https://www.afriscitech.com/en/news/on-schedule/1004-african-science-literacy-network-science-communication-and-journalism-workshop-holds-in-abuja

Waisbord, S., Shimp, L., Ogden, E.W., and Morry, C. (2010) 'Communication for polio eradication: improving the quality of communication programming through real-time monitoring and evaluation', *Journal of Health Communication*, 15: 9–24. doi: 10.1080/10810731003695375

WHO (2002) 'World Health Organization traditional medicine strategy 2002–2005', Geneva: World Health Organization.

Yankah, K. (1995) *Speaking for the Chief: Okyeame and the Politics of Akan Royal Oratory*, Bloomington, IN: Indiana University Press.

Zanna, U.A. (2019) 'Boosting science communication: network for journos, scientists launched in Nigeria', Yen Live, Available from: https://yenlive.com/news/index.php/news/2160-boosting-science-communication-network-for-journos,-scientists-launched-in-nigeria

7

Harnessing Indigenous Knowledge Systems for Socially Inclusive Science Communication: Working towards a 'Science for Us, with Us' Approach to Science Communication in the Global South

Konosoang Sobane, Wilfred Lunga, and Lebogang Setlhabane

Introduction

Recent trends in science communication have demonstrated that there is an increasing need for scientific information as well as the ability to access it. This has been especially true during the COVID-19 pandemic, where the extent of misinformation and disinformation (Ahinkora et al, 2020) has been a source of concern, with information-sharing as a public prerogative no longer monopolised by scientists and science communicators. The complexity of the current communication ecology is exacerbated by the diversity of available sources of information and the ever-increasing need to be first, right, and credible in sharing information. This era thus requires reflective thinking about the contextualisation of science communication epistemologies.

There has been increased appreciation of the fact that many scientific and social innovations that have the potential to empower society and facilitate social transformation can only achieve that aim through inclusive engagement methodologies and approaches. For example, Chivers and Hargreaves (2018) note that inclusive public engagement and participation methodologies are instrumental in realising socio-technical transitions. In the Global South, scientists and science communicators are increasingly acknowledging the significance of socially inclusive methodologies and approaches that will enhance participation in knowledge creation, knowledge-brokering, and science communication systems (Covello, 2021). Their transformative efforts are however hampered by the reality that most of the epistemologies and science communication insights are Eurocentric and fall short of being contextualised to Global South contexts. This lack of contextualised approaches results in the unequal distribution of and access to opportunities for effective public engagement with the processes and outcomes of science in the Global South.

The continued marginalisation of local insights through Eurocentric approaches to science communication calls for evidence-based advocacy for the advancement of socially inclusive approaches to science communication. In as much as the deficit model is often defined as one-way communication from scientists to the public without acknowledging other knowledge forms (Wibeck, 2013), this chapter argues that in marginalising the Indigenous knowledge systems in the Global South as a knowledge and communication base, the Eurocentric dominance of science communication in these regions is also in effect a problematic manifestation of the deficit model in practice.

Drawing from practical examples of science communication in the Global South, this chapter provides insights into how Eurocentric approaches to science communication, applied in these regions of the world, miss out on the opportunity to harness Indigenous knowledge systems. It then provides evidence-based examples of specific ways in which more contextualised approaches can be used and the value they would add to science uptake and appreciation. In particular, the chapter explores some of the key elements and practices of communication that can be harnessed to inform contextualised science communication and thus enhance inclusivity and co-creation in designing, implementing, and evaluating science communication and engagement in the Global South.

Science communication perspectives and practices: shifting paradigms from deficit models to public engagement

Traditional science communication perspectives and practices have been characterised by the persistence of the oversimplified deficit model in which communication is treated as a one-way stream whereby scientists or knowledge producers provide the publics with information intended to fill a knowledge gap (Wibeck, 2013). In this way, public audiences are treated as lacking relevant knowledge or experience and as not scientifically literate or interested in science (Simis et al, 2016). The point of departure for the deficit model is that 'deficits in public knowledge are the central culprit driving societal conflict over science'(Nisbet and Scheufele, 2009, p 1767). For that reason, low scientific literacy and lack of trust in science, as well as a lack of public understanding of science, are directly associated with deficits in scientific knowledge and are deemed to be easily remedied by disseminating knowledge to the public.

The challenge with the deficit model as applied in the Global South is that it misses out on the opportunity to harness the wealth of Indigenous knowledge systems that already exist locally and the in-depth contextualised understandings that these knowledge systems offer. Deficit model approaches to science communication also fail to harness the potential value of co-creation and collaboration with target audiences and the resulting empowerment of social actors to solve societal problems using scientific evidence (Mason and Mega, 2021; Scheufele et al, 2021). Co-created science communication approaches benefit from the core knowledge systems and practices of the target communities as well as existing experiential and contextual knowledge in these communities. They also benefit from harnessing the already inherent information-sharing tools, resources, and practices among the target audiences, which deficit models cannot do.

Exploring public engagement as an alternative

Given the deficiencies in the deficit model, and acknowledging the wealth of Indigenous knowledge systems that could inform behaviours and practices, science communication practices in the Global South are undergoing a paradigm shift towards participatory models with pillars of public participation and engagement as well as inclusivity. Recent literature (Alhassan et al, 2019; Weingart et al, 2021) shows that while the word 'engagement' seems to dominate policy and science communication discourses, clear definitions of key principles of engagement with science are lacking. There are also vague definitions of key concepts such as 'publics', 'citizen stakeholders', and 'non-scientists', which are often used very loosely in these discourses. This definitional fuzziness is even more glaring in the Global South context where it is widely known that there are 'sciences' and knowledge systems that are embedded and rooted in culture. But to what extent are these included in the definition of science and society? And to what extent are they incorporated into science communication and engagement activities?

Despite the ambiguity in the definitions of key concepts, there is a general understanding that the main objective of these participatory and engaging science communication practices is to provide opportunities for mutual learning between scientists and members of the public affected by science (Metcalfe, 2020). Such learnings include increased awareness of the cultural relevance of science and recognition of the importance of multiple perspectives and domains of knowledge to scientific endeavours.

According to Bauer and Jensen (2011, p 3), 'over the years, the term public engagement has taken the specific meaning of communicative action, to establish a dialogue between science and various publics'. In this way, the public is enabled to actively think about and become involved in science.

The public engagement approach often uses and builds on public understanding efforts while moving towards more comprehensive and interactive opportunities for dialogue and exchange. Through engagement, scientists and the public participate in discussions about the benefits and risks of science and technology impacting their daily lives. In doing so, questions and concerns can be better understood and addressed. Furthermore, involving a wide range of interested stakeholders can connect seemingly unrelated viewpoints, with potentially far-reaching effects. Public engagement with science is therefore seen to offer a more holistic, interactive approach that has the potential of getting people excited about science, increasing public trust in science, and embracing public attitudes and perceptions about science (Felt and Fochler, 2008).

Inclusivity as a pertinent characteristic of public engagement with science approaches: where are the gaps?

One of the characteristics of public engagement approaches to science communication is that they embrace inclusivity. Particularly for the Global South, where the communication ecology and participation in science systems are characterised by inherent inequalities, contextualising public engagement designs to give the public a voice is imperative. Inclusivity has, however, become very elusive, and science engagement continues to be driven by and approached through the lens of 'the scientist' (Simis et al, 2016). In addition, most of the public engagement approaches, frameworks, and epistemologies originate from the Global North (Weingart et al, 2021). This privileging of Eurocentric frameworks in Global South scenarios means that public engagement in these regions misses some of the critical actors, systems, and knowledge systems that exist in them. In addition, attempts to apply these Western-derived models and frameworks in the Global South further compromise inclusivity and become a barrier to the effectiveness of public engagement efforts, as it results in Indigenous communities and knowledge systems remaining at the margins of participation in knowledge production and access to science, as well as science communication, as noted by Finlay et al (2021).

One of the gaps rarely acknowledged in science communication discourses in the Global South is the persistence of the deficit model, which manifests itself in the adoption of Eurocentric approaches in contexts in which they do not adequately fit to the exclusion of contextualised local approaches.

The available literature (Seleti, 2010; Rasekoala, 2015; Ishinaha-Shinere, 2017; Finlay et al, 2021) has consistently asserted that there is a need to revisit the landscape of science communication in the Global South, re-envisioning not only epistemologies but also policies and practices. The lack of acknowledgement of the cultures that define local knowledge systems in the science communication system (Finlay et al, 2021), as well as widespread misunderstandings that lead to the trivialisation of these cultures, as observed by Seleti (2012), is a good example of this deficit model in practice. Seleti (2010) has long advocated for the mainstreaming of Indigenous knowledge systems within science policy frameworks on the African continent as a direct means of delivering emancipative Afrocentricity and epistemic liberation for African citizens and African languages in which these Indigenous knowledge systems are embedded. In fact, Ishinaha-Shinere (2017) remarks that this colonial thinking has become so systemically entrenched that science and technology (S&T) policies in the Global South will use justifications such as 'the shying-away of young people from S&T', 'accountability for research investment', and 'problem-solving on issues related to S&T and society' to cement traditional Western hegemonic science communication approaches in their systems. These disconcerting observations are consistent with earlier work by Palmer and Schibeci (2014), who observed the persistence of this same deficit model among the funding bodies that support research and science communication in the Global South. While there is an amplification of communication within the research community, Palmer and Schibeci (2014) established that there is less emphasis on communication with the broader community by these international funding organisations. This in itself is a demonstration of how the deficit model in the Global South is not only practised but also institutionalised and financially enabled to be sustainable. This chapter argues that this is a definition of the deficit model that is rarely acknowledged.

Science communication scholars agree that combatting the deficit model in the Global South requires that systems are opened up to accommodate a wide spectrum of cultures, knowledge systems, and practices. According to Finlay et al (2021), this entails creating spaces for reflective thinking and institutional practices. It also entails acknowledging the many creative practices, values, and knowledge systems that already exist in the Global South.

Multilayered exclusion factors that compromise inclusive science communication

Dimensions of exclusion, such as scientific literacy, the digital divide, and language barriers, continue to compromise the effectiveness and inclusivity of public engagement approaches to science communication. The inability of researchers to translate research into linguistically accessible formats that can be used to communicate and engage the non-researcher public compromises the potential for use and uptake of research and thus restricts its impact (Matias et al, 2021). An additional challenge is the continued neglect of the use of Indigenous languages and Indigenous communication methods, actors, tools, and platforms in public engagement as noted by Sobane et al (2021). This form of exclusion is more marked in multilingual contexts, where it has been proven that multilingual knowledge transfer facilitates improved public understanding and encourages the use of science in policy and practice. Kago and Cissé (2022) propose language harmonisation as one of the strategies that can be used to facilitate linguistic inclusivity in science engagement. There has been growing awareness that addressing inclusivity in science communication is crucial to ensuring that the knowledge that scientists and innovators invest in actually gets to different sectors of the population and thus has greater potential for impact. However, with all the different levels of exclusion remaining unaddressed in communication practices, the question of 'whose science and for whom' becomes glaring.

An additional layer of exclusivity is borne from the advent of digital communication, which has revolutionised science communication practices. Several studies identify the dialogic nature of digital platforms as an important feature in engaging different publics because it allows an exchange of views about the science and enables deliberations over the trustworthiness and applicability of science (Cahill and Ward, 2007; Wilcox, 2012). Digital science communication is therefore valued for its ability to facilitate the visibility of different voices in a dialogue where all voices are heard and valued, thereby closing the gaps between information-rich and information-poor publics (Jang et al, 2019). Digital communication has also been lauded for its potential to draw the attention of the public and keep them interested, engaged, and participatory in science-related matters (Park et al, 2020). Digital platforms are also seen as useful in the production of visualised tools that allow target audiences to have a better understanding of science, as well as enabling audiences of different levels of literacy to access and consume information (Bucchi and Saracino, 2016).

Although digital communication technologies have offered growing opportunities for science communication due to their cost-effectiveness and the ability to reach geographically disparate audiences almost simultaneously and for a lower cost burden (Bucchi and Trench, 2014; Lubinga and Sitto, 2021), internet access and affordability is one of the greatest barriers to inclusive access to scientific information (Okoth, 2022). Digital science communication often marginalises rural audiences and those whose socio-economic situation poses challenges of affordability in terms of access to digitised communication. In most cases in the Global South, these include rural-based knowledge producers and practitioners whom the digital divide bars from participating in specific digital science communication conversations. Rural-based science communication actors very often rely mostly on traditional media and basic digital connections such as text-based messaging as sources of information for engagement and participation.

While public engagement with science and its application seems to offer a more participatory inclusive approach, in the Global South there continues to be a need to have clearly defined mechanisms and epistemologies that account for *all* the publics. In particular, it is pertinent to have clearly defined participant roles to answer profound questions such as 'whose science and for whom?'

The increased marginalisation of Indigenous knowledge systems is another exclusionary factor in science communication. As noted by Finlay et al (2021), there is a dire need for science communication practices to be transformed to accommodate knowledges, practices, and systems often misunderstood and marginalised. Inclusivity in science communication needs to clearly account for the lenses and framings of Indigenous people and the existing knowledge that they hold. As Seleti (2010) observes, there is value in working towards the interfacing of Indigenous knowledge with other knowledge systems since they have a great deal of relevance and usability that can inform better processes of engagement. In particular, there is value in using the voices of Indigenous people in science communication to allow them to hear their stories being told in their voices and through their own experiences (Seleti, 2012). The value of Indigenous knowledge is further reiterated by Khumalo and Baloyi (2017), who show that these systems, which have sustained Indigenous communities for centuries before colonialism, were rendered underutilised by colonial practices and neglected, in contrast to the marked promotion of Western knowledge systems.

To overcome this exclusion, Finlay et al (2021) recommend transformed science communication systems and practices that acknowledge other knowledge systems in the knowledge economy, some of which have existed longer than Western knowledge, as noted by Rasekoala and Orthia (2020) and Seleti (2012). Of essential importance is the need to acknowledge the unique expertise, experience, and successful practices that have been built on ancient communication traditions (Rasekoala, 2015; Purnomo and Fauziah, 2018). In outlining some of the characteristics of this transformed science communication ecology, Rasekoala (2015) recognises citizen-centred approaches, co-creative joined leadership, and participatory approaches that involve social scientists and local actors as some of the enablers, pillars, drivers, and sustainers of this transformed landscape.

Inclusive science communication in the Global South: Indigenous knowledge good practice scenarios and exemplars

The emergence of digital communication technologies and the COVID-19 pandemic has made it imperative for science communication epistemologies and practices to lean more towards digital and online science communication. While these innovations and conventions may have altered interpersonal communication and hence science communication practices, some scholars (see, for example, Ayangunna and Oyewo (2014)) acknowledge that Indigenous communication systems still exist and that science communication can tap into these systems. In particular, Sobane et al (2021) note that there are often well-established community systems and practices of communication in the African context, for example, which can be harnessed to develop multi-sectoral engagements that will create awareness about science and enhance appreciation of its value in everyday use. In the following section, the chapter highlights some of the inclusive, impactful, and transformative innovations that Indigenous communication systems can offer to science communication epistemologies and practices across diverse regions of the Global South.

Available literature shows that even before the introduction of mass communication, there were Indigenous communication systems for information sharing across different societies. These systems were important in that they facilitated the preservation and adaptation of specific cultural information. According to Mundy and Compton (1991), these systems continue to exist alongside the mass media and digital communication technologies. The key characteristics of these systems are the multiplicity of voices and communication actors, as well as the diversity of platforms and languages that can be used. This chapter argues that, if these are incorporated into science communication epistemologies and practices, they have the potential to inform inclusive and contextualised forms of engagement.

The multiplicity of voices and communication actors

The involvement of diverse voices in the design, implementation, and evaluation of science communication enables it to be inclusive and accessible to a wider audience. As observed by Sobane et al (2020) in a study of COVID-19 communication in selected Southern and East African countries, several communication actors have taken up the communication of COVID-19 prevention messaging. These include creative artists in the entertainment industry (performance artists, singers, and comedians), language services companies, fine artists, and community media such as community radio and newspapers. Of particular importance is the existence of Indigenous communication actors who already have a trusted voice within their community. Such people include community religious leaders, traditional leadership, traditional healers, and midwives. These communication actors are rooted in the community-based contexts of communication and attuned to the cultural sensibilities of their communities. In addition, the way they repurpose and repackage messaging in different formats to enhance reach and accessibility for different population groups is informed by the contexts and knowledge of what works in that targeted audience.

In communication about the development of an irrigation tank in the dry plateau of the Deccan in Southern India, Baumgartner et al (2004) note that the involvement of an elderly former village headman, some farmers, a few boys, and the village teacher in communicating information about the project to outsiders created a strategy with a multiplicity of local contextualised voices with more potential for reach and impact. Actors such as these are important in that they are already trusted and depended on by the local community, as noted by Wang et al (2019).

Another example of the engagement of multiple actors in a science communication initiative is a project Dutta and Das (2016) conducted to establish factors affecting the communication practices and expectations of individuals living in the villages of Purulia, in rural Eastern India. They found that social embeddedness and co-designing communication tools with rural communities are key aspects of a successful communication strategy. The communication designers drew several science communication digital images, and the community made inputs into how the images could be made more culturally meaningful (that is, to communicate the exact/desired meanings) to the local Indigenous communities. This co-creative approach ensured that the designs were effectively contextualised and socially embedded in the community's beliefs and culture, thus creating better prospects of uptake. In particular, co-design fostered inclusivity while social embeddedness allowed communication design to emerge organically through embracing local knowledge.

A multiplicity of voices increases the potential reach and uptake of communication by diverse groups, since each voice may have a specific appeal to and potential influence on a particular section of the population. Communication through multiple voices also enables contextualisation and simplification of the information, which, in turn, allows the public to better interact and engage with the information.

Additionally, multiple voices enable communication in different languages, making messages accessible to speakers of those languages. Despite the widely acknowledged language diversity in the Global South, and the acknowledgement of this as a valuable resource for science communication, English is still the dominant language of science communication. Over the years, scholars have proposed several ways in which science can be made accessible to those who do not speak English. These include resources such as translanguaging (Makalela, 2016) and ad-hoc interpreting and translation (Fatahi et al, 2010) for face-to-face interactions, as well as translating documents into accessible languages (Sobane et al, 2020). As Márquez and Porras (2020) note, English has become a gatekeeper that prevents people from accessing and participating in scientific discourse, while also barring the multiple cultural interpretations of science. This compromises the effective translation of research into action as noted by Momen (2009).

Effective science communication design, implementation, and planning need to carefully consider and harness the multilingual characteristics of the societies in which they are embedded. As complementing voices that repurpose, repackage, and translate science, science communicators should adequately tap into language diversity and address the communication needs of those who cannot access the science in English and yet need to be scientifically aware and informed for science uptake.

The value of multimodality in enhancing the reach of science communication

Another identified characteristic of Indigenous communication that can benefit science communication and is already being harnessed effectively in some cases is multimodality (Burn and Kress, 2018), which refers to the use of a variety of communication methods, including writing, audio-visual products, and creative arts. These modes offer innovative means to capture the attention of different audiences and improve science uptake. The development of communication tools in local languages facilitates access to science for a majority of the population who are local language speakers. Scholars of Indigenous communication note that the use of interactive platforms and channels of communication opens up opportunities for co-creative learnings and better appreciation (Mundy and Compton, 1991). According to Etumnu and Fab-Ukozor (2021) communication in Indigenous communities is done through different modalities to amplify the reach. These include artistic performances and narrative approaches that translate science into products that can easily be consumed by users.

A recent example is the Ethiopian government's efforts to communicate policies and programmes that alleviate rural food insecurity (Nigussie, 2021). In parts of rural Ethiopia, such as in the Tigray region, communicators have started integrating folk media forms, elements of Aa'dar (oral poetry) and Goila (folk songs) for example, into food security communication. These have reportedly shown the highest potential for science uptake because of their edutainment characteristics. Each culture has its own forms: song, dance, puppetry, festivals, plays, storytelling, debates, proverbs, parades, and so on. If these are integrated into science communication practices, they have the potential to facilitate co-creation with affected communities, engender culture-sensitive science communication, and enhance the prospects for trust in and appreciation of science.

In Latin America and the Caribbean, for example, a range of science communication practices are embedded in the social order and Indigenous communication skills that allow communities to respond to different adverse risks in their everyday lives. Indigenous people have used their traditional knowledge to prepare for, cope with, and survive natural disasters for many years. In Honduras and Chile in South America and Haiti in the Caribbean, people respond to climate disasters through need-pooling (Postigo, 2021). In these contexts, culture and social relationships that facilitate need-pooling are found to be fundamental in governing risk recognition and disaster risk communication.

Some Latin American countries have built remote networks of scientists and science communicators that come together to create communication projects in different media platforms, inclusive of science museums, interactive science centres, natural history museums, environmental parks, zoos, botanical gardens, and aquariums. The focus is on regionally produced research, science policy, and science-related stories from the region using local languages (Weitkamp and Massarani, 2018). There is an acknowledgement that still more opportunities need to exist in terms of communicating science with regional relevance. The remote network offers quality and culturally relevant scientific information for non-scientific audiences in Latin America. From their different disciplines and concerns, science communicators intend to restore the value of science as a fundamental part of the Latin American human cultural heritage.

In a study exploring the integration of scientific and Indigenous knowledge to enhance the community's capacity for disaster risk reduction, Wang et al (2019) observed that in Haikou Village in the Ningxia Hui Autonomous Region of China, Indigenous communication tools are still prevalent and can easily be harnessed for science communication. Gongs, hand-actuated alarms, and oral notifications that can also be amplified by long-distance loudspeaker can be used alongside mobile phones to communicate to the public and send alert messages to distant ranges of 5–10 km during a disaster. By so doing, Wang et al (2019) show that Indigenous communication methods were optimised for effective information dissemination through the integration of new technologies. This created an effective community-based disaster risk reduction system and an effective communication system that can be repurposed for other science communication initiatives.

Conclusion

Given the multiple factors that affect access to science communication, this chapter argues that science communication epistemologies adapted from the Global North need to be rethought when being applied to the Global South. In particular, there should be careful consideration of the ways in which systems and policies can work towards the alleviation of the deficit model, which is manifested in the blind adoption of Eurocentric approaches in the Global South. Factors such as scientific literacy and interest and the ecology that affects them have to be taken into account in designing and implementing science communication in the Global South. Also important is a careful reflection on the communication seeking and sharing behaviours in targeted communities and the ways in which already existing communication ecologies can be harnessed to bridge the communication gap between science users and knowledge producers. While it is good to use the new technologies that in so many ways accelerate the reach of communication, there is value in establishing ways to manage the digital divide and enhance science communication. The Global South has a wealth of communication practices, tools, and platforms, as well as culturally oriented communication actors that can be harnessed to enhance science communication epistemologies for these regions.

References

Ahinkorah, B.O., Ameyaw, E.K., Hagan, J.E. Jr, Seidu, A.-A., and Schack, T. (2020) 'Rising above misinformation or fake news in Africa: another strategy to control COVID-19 spread', *Frontiers in Science Communication*, 5(45): 1–4.

Alhassan, R.K., Nketiah-Amponsah, E., Ayanore, M.A., Afaya, A., Salia, S.M., Milipaak, J. et al (2019) 'Impact of a bottom-up community engagement intervention on maternal and child health services utilization in Ghana: a cluster randomised trial', *BMC Public Health*, 19 (971): 1-11.

Ayangunna, J.A. and Oyewo, B.A. (2014) 'Indigenous communication, religion, and education as determinants of attitudes towards STI/HIV/AIDS education in Igando community, Lagos state, Nigeria', *African Journal of Social Work*, 4(1): 59–77.

Bauer, M.W., and Jensen, P. (2011) ,The mobilization of scientists for public engagement', *Public Understanding of Science*, 20(1): 3–11.

Baumgartner, R., Karanth, G.K., Aurora, G.S., and Ramaswamy, V. (2004) 'In dialogue with Indigenous knowledge: sharing research to promote empowerment of rural communities in India', in A. Bicker, P. Sillitoe, and J. Pottier (eds) *Investigating Local Knowledge, New Directions, New Approaches*, Aldershot: Ashgate, 207–23.

Bucchi, M. and Saracino, B. (2016) '"Visual science literacy": images and public understanding of science in the digital age', *Science Communication*, 38(6): 812–19.

Bucchi, M. and Trench, B. (eds) (2014) *Routledge Handbook of Public Communication of Science and Technology*, New York: Routledge.

Burn, A.N. and Kress, G. (2018) *Multimodality, Style and the Aesthetic: The Case of the Digital Werewolf*, New York: Routledge.

Cahill, J. and Ward, I. (2007) 'Old and new media: blogs in the third age of political communication', *Australian Journal of Communication*, 34(3): 1–21.

Chilvers, J., Pallett, H., and Hargreaves, T. (2018) 'Ecologies of participation in socio-technical change: the case of energy system transitions', *Energy Research & Social Science*, 42 : 199–210.

Covello, T.V. (2021) *Communicating in Risk, Crisis, and High Stress Situations: Evidence-Based Strategies and Practice*, New Jersey: The Institute of Electrical and Electronics Engineers.

Dutta, U. and Das, S. (2016) 'The digital divide at the margins: co-designing information solutions to address the needs of Indigenous populations of rural India', *Communication Design Quarterly Review*, 4(1): 36–48.

Etumnu, E.W. and Fab-Ukozor, N. (2021) 'Indigenous Communication and the Prospects for Survival in the Modern Era' in I. Nsude (ed) *African Communication System in the Era of Artificial Intelligence*, Enugu: Rhyce Kerex Publishers, pp 69–79.

Fatahi, N., Mattsson, B., Lundgren, S.M., and Hellström, M. (2010) 'Nurse radiographers' experiences of communication with patients who do not speak the native language', *Journal of Advanced Nursing*, 66(4): 774–83.

Felt, U. and Fochler, M. (2008) 'The bottom-up meanings of the concept of public participation in science and technology', *Science and Public Policy*, 35(7): 489–99.

Finlay, S.M., Raman, S., Rasekoala, E., Mignan, V., Dawson, E., Neeley, L. et al (2021) 'From the margins to the mainstream: deconstructing science communication as a White, Western paradigm', *Journal of Science Communication*, 20(1): C02 1–12.

Ishihara-Shineha, S. (2017) 'Persistence of the deficit model in Japan's science communication: analysis of white papers on science and technology', *East Asian Science, Technology and Society*, 11(3): 305–29.

Jones-Jang, S.M., Mortensen, T., and Liu, J. (2019) 'Does media literacy help identification of fake news? Information literacy helps, but other literacies don't', *American Behavioral Scientist*, 65(2): 371–388.

Kago, G. and Cissé , M. (2022) 'Using African Indigenous languages in science engagement to increase science trust', *Frontiers in Communication*, 6: 1–5.

Khumalo, N.B. and Baloyi, C. (2017) 'African Indigenous knowledge: an underutilised and neglected resource for development', *Library Philosophy and Practice (ejournal)*, Available from: https://core.ac.uk/download/pdf/188123787.pdf

Lubinga, E., and Sitto, K. (2021). Health communication in Africa. In Winston Mano and Viola Milton (eds). *Routledge Handbook of African Media and Communication Studies*. New York: Routledge, pp 217–233.

Makalela, L. (2016). 'Ubuntu translanguaging: an alternative framework for complex multilingual encounters', *Southern African Linguistics and Applied Language Studies*, 34(3): 187–96.

Márquez, M.C. and Porras, A.M. (2020) 'Science communication in multiple languages is critical to its effectiveness', *Frontiers in Communication*, 5(31): 1–5.

Mason, S. and Merga, M. (2021) 'Communicating research in academia and beyond: sources of self-efficacy for early career researchers', *Higher Education Research & Development*, 41(6): 2006–2019.

Matias, A., Dias, A., Gonçalves, C., Vicente, P.N., and Mena, A.L. (2021) 'Science communication for social inclusion: exploring science & art approaches', *Journal of Science Communication*, 20(2): 5–24.

Metcalfe, J. (2020) 'Chanting to the choir: the dialogical failure of antithetical climate change blogs', *Journal of Science Communication*, 19(2): 1–19.

Momen, H. (2009) 'Language and multilingualism in scientific communication', *Singapore Medical Journal*, 50(7): 654–56.

Mundy, P. and Compton, L. (1991) 'Indigenous communication and Indigenous knowledge', *Development Communication Report*, 74(3): 1–3.

Nigussie, H. (2021) 'Back to the village: integrating folk media into rural food security communication in Ethiopia', in H.S. Dunn, D. Moyo, W.O. Lesitaokana, and S.B. Barnabas (eds) *Re-imagining Communication in Africa and the Caribbean*, Basingstoke: Palgrave, pp 273–292.

Nisbet, M.C. and Scheufele, D.A. (2009) What's next for science communication? Promising directions and lingering distractions', *American Journal of Botany*, 96(10): 1767–1778.

Okoth, D. (2022) 'Reliable internet unavailable for 90 pct of poorest', *Scidev.net*, Available from: https://www.scidev.net/global/news/reliable-internet-unavailable-for-90-pct-of-poorest/

Palmer, S.E. and Schibeci, R.A. (2014) ‚What conceptions of science communication are espoused by science research funding bodies?', *Public Understanding of Science*, 23(5): 511–527.

Park, Y.J., Sang, Y., Lee, H., and Jones-Jang, S.M. (2020) The ontology of digital asset after death: policy complexities, suggestions and critique of digital platforms," *Digital Policy, Regulation and Governance*, 22(1): 1–14.

Postigo, J.C. (2021) 'Navigating capitalist expansion and climate change in pastoral social-ecological systems: impacts, vulnerability and decision-making', *Current Opinion in Environmental Sustainability*, 52: 68–74.

Purnomo, A.R. and A.N.M. Fauziah (2018) 'Promoting science communication skills in the form of oral presentation through pictorial analogy', *Journal of Physics: Conference Series*, 1006(2018): 1–7.

Rasekoala, E. (2015) 'Science communication in a post-2015 world: the nexus of transnational, multidisciplinary and sociocultural contexts', in B. Schiele, J., Le Marec and P. Baranger (eds), *Science Communication Today – 2015: Current strategies and means of action*, Paris: Universitaires De Lorraine, pp 39–45.

Rasekoala, E. and Orthia, L. (2020) 'Anti-racist science communication starts with recognising its globally diverse historical footprint', Available at: https://blogs.lse.ac.uk/impactofsocialsciences/2020/07/01/anti-racist-science-communication-starts-with-recognising-its-globally-diverse-historical-footprint/

Scheufele, D.A., Krause, N.M., Freiling, I., and Brossard, D. (2021) 'What we know about effective public engagement on CRISPR and beyond', *Proceedings of the National Academy of Sciences*, 118(22): 1–8. Available from: https://www.pnas.org/doi/epdf/10.1073/pnas.2004835117

Seleti, Y. (2010) 'Interfacing Indigenous knowledge with other knowledge systems in the knowledge economy: the South African case'. Library Symposium 2010*: Presidential Meeting. Knowing is not enough: engaging in the knowledge economy*. Stellenbosch University. 18–19 February 2010, Available at: http://scholar.sun.ac.za/bitstream/handle/10019.1/396/seleti_knowing_2010.pdf?sequence=2

Seleti, Y. (2012) 'The value of Indigenous knowledge systems in the 21st century', in J.K. Gilbert and S.M. Stocklmayer (eds) *Communication and Engagement with Science and Technology*, New York: Routledge, 267–78.

Simis, M.J., Madden, H., Cacciatore, M.A., and Yeo, S.K. (2016) 'The lure of rationality: why does the deficit model persist in science communication?', *Public Understanding of Science*, 25(4): 400–14.

Sobane, K., Nyaga, S., Ssentanda, M., and Ntlangula, M. (2020) 'Using multi-voiced and creative approaches to enhance COVID-19 messaging: learning from east and southern Africa', *HSRC Review*, 18(3): 43–4.

Sobane, K., Riba, I., and Lunga, W. (2021) 'The value of the fusion of Indigenous and contemporary knowledge in developing communication as interventions to reduce teenage pregnancy in South Africa', in B. Falade and M. Murire (eds) *Health Communication and Disease in Africa*, Singapore: Palgrave Macmillan, pp 333–345.

Wang, Z., Liu, J., Xu, N., Fan, C., Fan, Y., He, S. et al (2019) 'The role of Indigenous knowledge in integrating scientific and Indigenous knowledge for community-based disaster risk reduction: a case of Haikou Village in Ningxia, China', *International Journal of Disaster Risk Reduction*, 41: 1-9.

Weingart, P., Joubert, M., and Connoway, K. (2021) 'Public engagement with science -origins, motives and impact in academic literature and science policy', *PloS ONE*, 16(7): 1–30.

Weitkamp, E. and Massarani, L. (2018) 'Branching out: new JCOM América Latina for dynamic science communication community', *JCOM: Journal of Science Communication*, 17(2): E.

Wibeck, V. (2013) 'Enhancing learning, communication and public engagement about climate change – some lessons from recent literature', *Environmental Education Research*, 20(3): 387–411.

Wilcox, C. (2012) 'Guest editorial: it's time to e-volve: taking responsibility for science communication in a digital age', *The Biological Bulletin*, 222 (2): 85–87.

8

Indigenous Science Discourse in the Mainstream: The Case of 'Mātauranga and Science' in *New Zealand Science Review*

Ocean Ripeka Mercier and Anne-Marie Jackson

Introduction

Since 2005, researchers and scientists in Aotearoa New Zealand who seek public funding for their work have been encouraged to situate their research in relation to the government's Vision Mātauranga (VM) policy. VM seeks to 'unlock the science and innovation potential of Māori knowledge, resources and people' through investment in Māori knowledge, people, resources, partnerships, and improved relations between Māori people who are indigenous to Aotearoa New Zealand, and the British Crown, represented by the New Zealand government (Ministry of Research Science and Technology, 2007). VM builds from the premise that 'Māori success is New Zealand's success', and the policy incentivises researchers to work with and engage with mātauranga Māori (Māori knowledge) in the Aotearoa New Zealand research, science, and innovation (RSI) sector. VM also supports those including mātauranga (Māori knowledge) within public outreach and science communication. Today, there is an expectation that public conferences and seminar series will include sessions on mātauranga Māori. Museum displays, such as the permanent Te Taiao exhibition at Te Papa Tongarewa: Museum of New Zealand, engage both science and mātauranga. National broadcasting imperatives to produce media in te reo Māori (the Māori language) along with VM policy saw the funding, production, broadcast, and international distribution of the television show *Project Mātauranga* (Douglas 2012; 2013), which 'celebrates Māori innovation in the science sector'. Each half-hour programme showcases how mātauranga Māori is working alongside Western science (Mercier et al, 2014, p 70). Collaboration is not trivial, and many have written about the distinctions between mātauranga and science (Roberts, 1996; Jackson and Mercier, 2020; Morgan and Manuel, 2020) and explored Māori philosophy's uniqueness (Mika, 2012; Jackson, 2013; Stewart, 2021a).

The New Zealand education sector and education outreach initiatives have engaged both mātauranga and science for decades: in curriculum design for Māori language immersion and bilingual schools, within schools and universities, tertiary outreach to primary and secondary students, and cross-sector initiatives designed to 'inspire Māori and Pacific students to see themselves as scientists' (Mercier van Berkel, 2021, p 231). The National Institute of Water and Atmospheric Research (NIWA) Wellington Regional Science Fair has for nearly a decade offered prizes to students whose experiments or exhibits engage te reo Māori and mātauranga Māori. From 2024 onwards, the mainstream secondary school science curriculum is expected to include mātauranga. Initiatives to promote science capacity building among Māori students include university scholarships and internships, with the Pūhoro Academy, for example, prioritising 'STEMM', or science, technology, engineering, mathematics, and mātauranga. Herein lies the unique opportunity of Aotearoa New Zealand, that the Indigenous knowledge traditions form a unique and dynamic ontology and contribute to developing society alongside science. However, while VM policy has resulted in some increased capacity of Māori and visibility of mātauranga Māori in the RSI sector, in practice non-Māori researchers often lack the understanding or capability to engage productively with Māori. This has resulted in consulting rather than collaborating with Māori, 'tick boxing', the token addition of a Māori voice or 'flourish' to projects, treating mātauranga as 'another data source', and other extractive practices. Weak understanding leads science communicators to exclude mātauranga, or invalidate mātauranga, and by extension Māori people as rational thinkers. What is not stated in VM policy, or recognised in these engagements, is that mātauranga Māori itself is still subject to governance by Māori. Mātauranga has unique Māori practices by which it is governed, anchored in the Pacific and steeped in hundreds of years of history in Aotearoa.

This contribution describes a collaboration with *New Zealand Science Review* (NZSR) and shines light on how we, as Māori women researchers, experienced the communication of Indigenous science in this forum, as an example of science communication more broadly. This case reveals compatibility and tension between different modes of 'knowledge governance' (van Kerkhoff and Pilbeam, 2017). We revisit and analyse the emails, editorial discourse, and critical reception of these special issues. What mātauranga governance practices does the form of a special journal issue allow and exclude? Through this discussion, we highlight ongoing European dominance and identify key barriers to epistemic liberation of Māori knowledge within traditional academic scientific discourse in New Zealand. But we also suggest alleviating these impacts by recognising, anticipating, and accounting for the different knowledge governance practices. We discuss in detail two examples of governance dissonance. In the first, damaging critiques of our work were published without our knowledge or permission. In the second, mātauranga was disentangled from the people connected to it and treated as a tradable commodity. We suggest that science governance systems need to view mātauranga Māori as an independently governed system with some mātauranga under Crown care. This must be recognised for Māori people to advance towards equitable and innovative participation in science.

Knowledge governance

Mātauranga Māori (Māori knowings, knowledges, and knowledge-producing methods) has a 700-year history of development, dating from the first Polynesian migrants to Aotearoa. Their knowledge base expanded over generations, from a Pacific Island foundation to reflect the radically different landscape and ecosystems of Aotearoa's islands. Despite the Indigenous pedigree of science in this country (Hikuroa, 2016), the current knowledge governance system is based on the British European knowledge system that arrived in the late 18th century. This may not be surprising given the Treaty of Waitangi (ToW) allowed the Crown to assume government over the British in New Zealand. The Crown also promised, in Article 2 of Te Tiriti (the Māori version of ToW), to protect the right of Māori to their 'taonga katoa' (all treasures, precious resources). Taonga usually refers to natural resources, or works derived from natural resources, but also includes mātauranga (for example Te Aho, 2021). The Waitangi Tribunal report into the claim of treaty breaches related to native flora and fauna, filed as Wai 262, found that the Crown failed to protect, and actively suppressed, taonga such as te reo and mātauranga (Waitangi Tribunal, 2011). Directly, through legislating against mātauranga Māori (Tohunga Suppression Act 1907) and indirectly, through legislation (such as the Land Drainage Act 1908) that removed environments that were taonga-rich natural repositories of mātauranga Māori.

Cornell's *Why, What, How* framework (2021), applied to knowledge governance in Aotearoa New Zealand, invites us to suggest seven key aspects of knowledge governance that may help to understand the processes, systems, and controls to safeguard and grow resources within any knowledge governance system:

1. expertise and authorities;
2. epistemological, ontological, and philosophical basis;

3. data, information, knowledge, understanding, wisdom;
4. research institutions and knowledge repositories;
5. knowledge transmission and education;
6. communication and connection with publics;
7. economies of application.

The agents, institutions, and processes in the dominant knowledge governance system are probably easy for the reader to recognise and name, looking at the seven aspects in the list, so they won't be rehearsed here. It is worth noting, however, that the dominant knowledge system is not often examined in this way by practitioners within, so it is a worthwhile exercise to reveal the pervasive and far-reaching ways that our knowledge is curated and governed. Mātauranga Māori includes all of these seven aspects – configured and played out differently but nonetheless provoking recognition of mātauranga as a governed knowledge system. Importantly, these seven aspects are intertwined, reasserting how inappropriate it is to simply extract one aspect – item 3, the data – from a functioning, autonomous knowledge governance system. Mason Durie also warns that the metrics and scales of one knowledge system should not be applied to the other (Durie, 2005). We will refer back to this typology as we discuss some of the issues that arose in our communication of mātauranga to an audience of people largely operating according to the mainstream knowledge governance system.

The *New Zealand Science Review* invitation

In 2018, the editor of NZSR, a peer-reviewed journal of the New Zealand Association of Scientists, sought to address 'an acknowledged void in the understanding of many in the present research community' regarding mātauranga and VM policy. The editor invited me (Dr Ocean Mercier) to submit an article that could shed light on this, which I did, and 'Mātauranga and science' was published (2018). The reviews revealed an awkward smashing of toes at different epistemological and relational standpoints. The editor mediated this by acknowledging "there is a degree of 'emotion' in the comments offer [*sic*] by the reviewers. This however, I suggest, is to be expected in this area and not a reflection on you personally". Uncertainty and pushback from reviewers to points made in Ocean's article unexpectedly revealed a 'void in understanding' among science communication leaders (Editor to Ocean, 23 July 2018). Upon the advice of a Māori colleague, I suggested to the NZSR's editor that my expert peers in mātauranga contribute to further conversation through a dedicated issue of NZSR. This would explore the issues further, raise awareness of the breadth of work occurring, and reveal the many ways that mātauranga contributes to different disciplines.

A whole issue on 'mātauranga and science' was met with enthusiasm by the NZSR. I drafted a call for proposals and a list of potential contributors and was given the go-ahead from the editor and editorial team of the NZSR. I reached out to Dr Anne-Marie Jackson of University of Otago and Dr Pauline Harris of Victoria University of Wellington to join me as co-editors. Working with Anne-Marie and Pauline (who supported us as a co-editor in the early stages, and was involved later as a co-author on one of the published articles) strengthened the work, as summed up in the whakatauki (proverb) 'Ehara taku toa i te toa takitahi' (My strength is not in the strength of me alone), contrasting with my lonely experience as sole author of 'Mātauranga and science' (2018).

We received 19 expressions of interest, and a total of 15 article submissions, ranging in length from 3,000 to 9,000 words. Between 2019 and 2020, we curated and compiled two special issues of the NZSR, entitled 'Mātauranga and science in practice'. The work of 100 contributors, mostly Māori, across 12 original peer-reviewed contributions (one in two parts), two forewords, and two introductions was published in print and online across two issues in 2019 and 2020. Topics included technology, education, philosophy, Indigenous science, demography, marine science, astronomy, environmental science, geospatial tools, and policy, reflecting the variety and depth of Māori-led research.

We invited thought leaders in the RSI sector to contribute a foreword to each issue: the first was co-authored by the Prime Minister's Chief Science Advisor Professor Juliet Gerrard and Forum of Chief Science Advisors member Professor Tahu Kukutai (Gerrard and Kukutai, 2019). The second (Hutchings and Martin, 2020) was co-authored by Drs Jessica Hutchings and Willy-John Martin, co-chairs of Rauika Māngai, the Māori oversight group for the National Science Challenges (a ten-year investment in 11 science themes driven by public interest). The latter foreword coincided closely with Rauika Māngai's release of *A Guide to Vision Mātauranga: Lessons from the New Zealand Science Sector* (2020), which brought together many Māori research voices reflecting on the VM policy.

Mātauranga governance practice: our unspoken rules

We wanted to create a space for us to discuss our research with, by, and about ourselves as Māori, in line with Kaupapa Māori principles (see Smith, 1999) and anticolonial strategies (Simpson, 2004).

Kaupapa Māori theory principles are: tino rangatiratanga (the self-determination principle); taonga tuku iho (the cultural aspirations principle); ako Māori (the culturally preferred pedagogy principle); kia piki ake i nga raruraru o te kāinga (the socio-economic mediation principle); whānau (the extended family structure principle); kaupapa (the collective philosophy principle); Te Tiriti o Waitangi (Treaty of Waitangi) (Smith, 1997; Smith, 1999); and āta (growing respectful relationships) (Pohatu, 2005).

All of the principles are of importance, and in the context of knowledge governance considerations for appropriate science communication in Indigenous contexts we used taonga tuku iho (the cultural aspirations principle) and tino rangatiratanga (the self-determination principle).

In relation to taonga tuku iho (the cultural aspirations principle), we applied two tikanga (principles): mana (authority, integrity) and whānau (kinship). We felt privileged to have access to the journal and to open up a predominantly science space to the discourses and norms of our Māori peers working at the interface between mātauranga and science. We brought mana and whānau principles to our work with the journal also.

As co-editors, we are PhD graduates, researchers, writers, professors and associate professors, heads of academic units, research programme leaders, investors of external research income, we contribute to research and policy environments, and we're respected and networked with many other researchers – Māori and non-Māori – in the New Zealand RSI sector. Despite its challenges to us as Māori in science (McAllister et al, 2020) and more particularly Māori women in science (McKinley, 2008), we have nonetheless crafted successful careers in the dominant knowledge system and have laid down pathways for others. We are also respected in our iwi (tribe) and hapū (subtribe) communities for our education, ability to teach others, knowledge of te reo Māori (Māori language) and mātauranga, and for our advocacy voices. We understand that there are unspoken (and unwritten) governance rules for mātauranga and, as a result, conduct ourselves in unique ways at our institutions.

An important aspect of taonga tuku iho is how our taonga – including mātauranga – are governed, cared for, managed, maintained, and enabled to flourish, now and for future generations. As leaders, kaitiaki (guardians) of mātauranga, we are tasked with safeguarding mātauranga as a taonga resource; being especially mindful and careful of whether and how we share and communicate mātauranga within the dominant science-governed system. The journal NZSR became something of a microcosm, focusing, reflecting, and illuminating some of the difficulties that the wider system grapples with when communicating science from dual traditions. We're also kaitiaki of our *relationships* with others – existing and new.

Through asserting our tino rangatiratanga (the self-determination principle), we aspired to make the journal reflect us in content and design. We wanted to use te reo (Māori language) where possible. Contributors could submit in te reo Māori, and we later invited all our contributors to submit the abstracts of their reviewed and accepted publications in te reo Māori also. Some did not have te reo Māori ability or capacity, so we employed a talented te reo graduate, Eru Kapa-Kingi, to translate these.

We employed learnings from Simpson's (2004) anticolonial work, focusing on the 'recovery of Indigenous intellectual traditions, Indigenous control over Indigenous national territories, the protection of Indigenous lands from environmental destruction, and educational opportunities that are anticolonial in their political orientation and firmly rooted in traditions of their nations' (2004, p 381). Every article in the special issue focused on Indigenous knowledge, all written by Indigenous scholars, many with community members, with non-Indigenous allies, all on topics of relevance and importance for Indigenous peoples.

Mātauranga and science governance practices: alignment and contrast

As experienced researchers, we are no strangers to working within the governance systems around journal publication in the dominant knowledge governance system and have the additional experience of also navigating and negotiating our principles as Māori researchers. This often forces us to find creative ways of working around processes and practices so we can also ensure and uphold mana and integrity of knowledge, knowings, and knowers. But this is not a smooth negotiation, as we are ultimately answerable to two or more masters. Revisiting the governance typology highlights aspects where boundaries had to be negotiated but were relatively smooth. The exercise also shows areas in which it was less within our control to maintain mātauranga governance principles while satisfying mainstream processes:

1. We offered a platform (the journal) to mātauranga science *experts and authorities* wishing to publish. This reflected tino rangatiratanga (the self-determination principle) for the experts, to publish on topics of importance to their communities, and to satisfy other authorities and decision-makers of the value of the new knowledge. The verifying authorities had to have like knowledge and understanding, so our pool of reviewers was often quite select. Issues subsequently arose when the legitimacy of knowledge published in our special issues was criticised by non-experts.
2. The *epistemological, ontological, and philosophical basis* here was a blend of mātauranga and science, and the review process needed to verify the originality and worth of the new knowledge on both of those bases, which is an example of taonga tuku iho (cultural aspirations principle). This included expressions of science in te reo Māori. A Māori colleague expressed some reservations about whether te reo Māori should be included – would it be sufficiently valued in this particular journal? These turned out to be valid concerns.

3. We curated substantial new *knowledge* generated by authors – *understanding and wisdom* came about through connection and making the *data and information* relevant to the mātauranga and science setting, highlighting kaupapa (the collective philosophy principle) whereby all authors contributed to the shared vision of the journal.
4. The NZSR is a *knowledge repository* we wanted more exposure in – we expected an element of autonomy over our special issues while being committed to uphold the standards of the New Zealand Association of Scientists and the journal itself. However, some *research institutional* norms competed against this *mana* (see subsequent discussion), and this disruption revealed the need for Simpson's (2004) approach to anticolonialism.
5. We achieved *transmission and education* – the original invitation arose from an acknowledged gap in scientists' knowledge regarding mātauranga and science. The special issue generated an educational resource to enhance the mātauranga literacy of the NZSR readers, as well as support Māori aims to record and disseminate our stretchy work at the mātauranga–science interface. See, for example, the G.T. Stewart articles 'Mātauranga and Pūtaiao: the question of "Māori science"' (2019) and the Kepa Morgan and Robyn Manuel (2020) article 'Western science and Indigenous wisdom: is integration possible, practical, plausible?'
6. *Communication and connection with publics* – there is increasing public appetite to learn more about mātauranga, as this helps people connect with local lands to develop an identity germane to Aotearoa. This has led to increased inclusion of mātauranga in research and science communication. However, public discourse about mātauranga can become racialised and polarised by science communicators who proclaim the primacy of modern science. Self-assumed experts on mātauranga and Indigenous knowledge systems speak in ways that lead to damaging and misleading forms of science communication, destabilising moves towards equity and diversity (see subsequent discussion).
7. *Economies* – a Ngā Pae o te Māramatanga (Ngā Pae) grant allowed us to disseminate the NZSR freely to non-subscribers. However, subsequently parts of the work were treated as non-relational commodities, extracted, and reprinted.

With financial support from Ngā Pae, we printed and mailed more than 300 copies of each issue to non-subscribers. We distributed them to students and assigned readings in undergraduate courses at Victoria University of Wellington and postgraduate offerings at the University of Otago within Anne-Marie's co-led research group, Te Koronga. We have also distributed them at iwi gatherings, Crown Research Institutes, and educators' fora. The NZSR journal's usual practice is to keep their latest issue behind a paywall until the next issue is released, but our first issue was made available freely online from its release. From the perspective of whānau and mana over the curated collection, we appreciated this arrangement, which gave contributors immediate access to their and their peers' manuscripts. We hoped this widespread dissemination of a verified evidence base would strengthen people's confidence to communicate mātauranga science.

Mātauranga subjected to epistemological scrutiny

An article about Te Koronga (Jackson et al, 2019) in our special issue attracted a written letter to the editor from a physicist who emailed it to Anne-Marie and then to the NZSR editor for publication. His letter expressed dismay at the presence of 'myths' in Indigenous science curricula. Among other things, he claimed that 'science is universal' and denied any boundaries between knowledges:

> Science cannot be predicated on mythical cosmologies as is proposed in the description. Science is universal. Myths are local. There is no such thing as Western science or Maori science. Empirical science transcends all boundaries. It is totally inclusive in its philosophical constructs and its approach to collecting and validating data. (Tallon, 2020)

The writer went on to argue that 'universal science' is inclusive, a 'science for everyone':

> The basis under which science is to be communicated to our diverse communities is not through our individual myths. Science does not need that. It has its own compelling dynamic – its own imperative. Science speaks to everyone through the understanding it creates, through its sheer wonder and through the amazing technologies it dispenses. Best of all, science promotes itself through the passion of its agents – scientists. These are the ones who must be harnessed to achieve the goal of strengthening Maori engagement in science and scientific research, and they are very willing. (Tallon, 2020)

We were informed that the letter had travelled beyond Anne-Marie's inbox, to the point that it had been copyedited for publication in the NZSR. We were offered the option to reply or not: 'In my experience a short rather than a long reply serves the purpose' (Editor to us both, 8 September 2020).

Strict adherence to a modernist position on science is probably not typical of the 21st-century New Zealand scientist. However, these views went on to receive more airing when the author resubmitted his letter to the *Chemistry in New Zealand* journal. Independently, a letter in a similar vein was submitted to the editor of the *New Zealand Listener*, co-signed by seven professors and emeriti at the University of Auckland. It was published by the weekly current affairs magazine and entitled 'In defence of science'. In response, a philosopher asked 'what exactly does science need defending from?' noting further that '[t]he academic disciplines have defined themselves by such acts of intellectual exclusion, and the boundary between science and Indigenous knowledge has historically helped science to define itself' (Stewart, 2021b). The letter's authors noted what they see as 'disturbing misunderstandings' emerging in the new secondary school science curriculum on the colonial origins of modern science. The letter also stated that 'Indigenous knowledge may indeed help advance scientific knowledge in some ways, but it is not science' (Clements et al, 2021).

Responses appeared in defence of mātauranga. The New Zealand Association of Scientists published a statement on the letter:

> Specifically, like many others we were dismayed to see a number of prominent academics publicly questioning the value of mātauranga to science in the Listener this week, and claiming that science does not colonise. While as scientists we clearly see the value of science and the good it can do, we must also acknowledge that science has an ongoing history of colonising when it speaks over Indigenous voices, ignores Indigenous knowledge, and privileges a limited, Western-dominated view of science. (New Zealand Association of Scientists, 2021)

An open letter led by Professor Shaun Hendy and Associate Professor Siouxsie Wiles from Te Pūnaha Matatini, the Centre of Research Excellence in Complex Network Systems, gathered hundreds of signatures in support. In the following quotation, the authors link the recurring claim of 'universalism' in science to exclusion and lack of diversity – which rebuts the two arguments cited earlier: 'And while the Professors describe science as "universal", they fail to acknowledge that science has long excluded Indigenous peoples from participation, preferring them as subjects for study and exploitation. Diminishing the role of Indigenous knowledge systems is simply another tool for exclusion and exploitation' (Hendy and Wiles, 2021). An author whose article in the NZSR was a target of the Tallon letter, and whose curriculum work was targeted by the 'Auckland 7', analysed and critiqued these letters to the editor as 'word weapons' (Stewart, 2021c). Both letters to editors explicitly laid out the authors' academic credentials, claiming expertise while speaking into an area of non-expertise (Stewart, 2021c). They also revealed their philosophical and sociological knowledge deficit: '[M]ost or all of the seven professors were making judgemental claims about Māori knowledge that were well outside their professional brief – but the second, more interesting, question to ask is: What do they actually know about science?'

(Stewart, 2021b). These ungoverned attacks on another knowledge system reveal the anxiety some feel about non-Western knowledge governance systems. Tina Ngata (2021) remarks that the outbursts are 'bold evidence for the endurance of White supremacy within academia and science, a feeling of threat in the face of the colony's eroding power'.

Discussion, conclusion, and implications

What do these issues mean for inclusive science communication in New Zealand, across various platforms? Two problems immediately evident are a science philosophy knowledge deficit, and a capability gap in understanding other ways of knowing. Scientists need to make room for mātauranga in the ways they talk about science. This aligns with Simpson's approach, whereby scientists must be willing to 'step outside of their privileged position and challenge research that conforms to the guidelines outlined by the colonial power structure and root their work in the politics of decolonisation and anti-colonialism' (Simpson, 2004, p 281).

There is room for mātauranga in many museums in the country, and it would underpin certain exhibits, for example the science of taonga such as Māori cloaks. Mātauranga is a recurring and normalised feature within science communication in public science festivals, seminars, and conferences. The 2021 New Zealand International Science Festival coincided with Matariki, the Māori celebration of the New Year with the rising of Pleiades, and workshops about Māori lunar and astronomical knowledge were offered. Most New Zealand environmental conferences now programme a session of talks on mātauranga. The National Science Challenge programmes were set up in 2014 to support public engagement in science, and the 11 thematic areas (for instance Our Biological Heritage, Sustainable Seas, Ageing Well, and so on) were determined by public submission and vote. They are another key arena in which mātauranga and Māori researchers collaborate with science at all levels, including co-leadership and governance. In National Science Challenge, BioHeritage public outreach events – such as the Crazy and Ambitious conference – interactive exhibits and seminars show that old mātauranga revived and new mātauranga produced has a prominent place in science discourse. Mātauranga has become widespread in the science communication mainstream in New Zealand.

This communication does not enter a Māori space nor a neutral space, however, and the seemingly ever-present negative discourse about mātauranga – even if it appears to be coming from the margins – continues to distract and dilute our efforts to diversify and grow mātauranga in the sector. The rise of mātauranga needs to run parallel with the strengthening of Māori people's capability, so that ancestral knowledge inheritors are given space in the discourse and retain rights to decide how and where their mātauranga is accessed, applied, used, developed, and created. This is a governance issue, and these decisions about mātauranga need to be made by and for Māori, according to Kaupapa Māori principles. The same applies for Indigenous and non-mainstream knowledges everywhere.

In staking out space within a national science journal, we expected to improve science communication on multiple fronts, but perhaps primarily through giving scientists the evidence of effective collaboration across knowledges. Science communication here is not just scientists speaking to publics but us whispering to scientists in the background, prompting them upon what should be said (or not), by whom, when, and how. By reporting to scientists, through the NZSR, about how mātauranga and science are working together, we were addressing a knowledge gap in one of the country's key science publications. We aimed to diversify the science knowledge base and shift understandings about what science is. We sought to affirm Māori working in the sciences and to educate the scientific community. We expected that publishing mātauranga in a mainstream science journal would enhance the trust of scientists in this discourse, thus their confidence in it, and willingness to make space for Māori and for mātauranga.

In Māori tradition, naming something implied you had a physical connection to the land and had done a reflection and assessment of the character of the land in order to understand its essence. *Communication* through naming is a practice of enacting *governance* rights. Applied to knowledge, it has been a long-suffered injustice that Māori see their mātauranga 'named and claimed' into written texts by generations of non-Māori natural scientists, anthropologists, ethnographers, and the like. Our knowledge systems are intrinsically valid and don't need publication to give them mana within their own systems. While publication in mainstream journals may give mātauranga more mana in the eyes of the mainstream, what is undeniable is that mātauranga is governed, and Māori must continue to govern mātauranga, in Māori ways, wherever mātauranga is homed.

A governance lens helps to identify how we can have continued sovereignty over our knowledge systems and their communication. The typology suggested here, of seven aspects to knowledge governance – whether the dominant science knowledge system, an Indigenous knowledge system, or mātauranga Māori – asserts in new terms the inappropriateness of measures that isolate Indigenous data and extract it for the great public archive. A knowledge governance system lens also helps to identify and recommend specific strategies for engagement by the dominant knowledge governance system with Indigenous knowledge governance systems, and vice versa.

Here we've noted that a key audience for communication about Indigenous science is scientists themselves, as their voice is enduringly influential in Aotearoa New Zealand, even as the country carves out a unique knowledge system built on dual traditions. Specific shifts in publishing and communication could achieve broader acceptance of mātauranga, and mātauranga as a governed system. Rules by which to engage with Māori people and their mātauranga must move beyond extractive approaches if we are to avoid the kinds of 'special issues' that arose here when traditional science governance practices breached the governance boundaries of mātauranga. We were not expecting to have to undo Eurocentric legacies in research publishing. Through centring te reo Māori and mātauranga Māori in a scientific forum, these legacies found and confronted us. Fortunately, as quoted excerpts show, this is counter-balanced by overwhelming support for mātauranga at the NZSR and among New Zealand scientists.

Publishing was a vehicle through which we sought to give tino rangatiratanga and mana to our participants and their communities, their knowledge traditions, and their research. We sought to honour and enhance our research relationships by bringing our community of researchers to the pages of the journal together, as a whānau. These were forms of governance that helped realise our aspirations for epistemic liberation. Explicitly stating our modes of knowledge governance might deflect future issues that bring Eurocentric science governance into conflict with our own.

There are several lessons that can be drawn upon from our experiences and theoretical positioning that are of relevance for Indigenous peoples, scholars, activists, community members, advocates, and students worldwide. For example:

- The wider positioning of this work is within the global Indigenisation of the world whereby discourses of conscientisation, resistance, and transformative action (Smith, 1997; Smith, 1999) are critical tenets in the collective movement of Indigenous development. We hope our shared experiences engender a continued sense of the collective contribution to the transformative nature of our Indigeneity.
- This triple approach we have drawn upon of contextualising Cornell's *Why, What, How* framework (2021) within an Indigenous theoretical grounding (Smith, 1997; Smith, 1999) and Simpson's (2004) anticolonial framing is a useful approach to advance the theoretical underpinnings of science communication in relation to knowledge governance. Indeed, this wider approach can be applied within any Indigenous context whereby the grounding is framed in the specific location to open a continuous dialogue and intersectionality.
- The ways in which we have engaged in science communication, and extended this to knowledge governance, have been to privilege Indigenous voices, and uphold cultural practices within an anticolonial agenda, through disrupting the colonial powers of knowledge governance and knowledge production. This can be translated into contexts that are by Indigenous, for Indigenous; at the interface; and within non-Indigenous

spaces with a decolonising ethic. The work we have shared in the special issues provides examples across these three contexts and highlights that these lessons are non-linear, are dynamic, and both unpick and recreate an interwoven understanding of the world, as Marsden states 'the universe is not static but is a stream of processes and events' (Marsden, 2003, p 21).

Finally, we conclude with a quote from Marsden that summarises the paradoxical nature of the challenges of being Indigenous and the need to exist within another set of principles or worldviews, such as we have described in communicating Indigenous knowings within a Western science system: 'Abstract rational thought and empirical methods cannot grasp the concrete act of existing which is fragmentary, paradoxical and incomplete. The only way lies through a passionate, inward subjective approach' (Marsden, 2003, pp 22–3).

Acknowledgements

We acknowledge funding that enabled this work from Ngā Pae o te Māramatanga, Te Koronga, and Manaaki Whenua Landcare Research. We thank contributors to the *New Zealand Science Review: Mātauranga and Science*, particularly those whose experience we've referenced. We thank Allen Petrey, the editor of the *New Zealand Science Review*, for the opportunity to be guest editors. Finally, we thank our reviewers pre- and post-submission: your feedback greatly improved this chapter.

References

Clements, K., Cooper, G., Corballis, M., Elliffe, D., Nola, R., Rata, E. et al (2021) 'In defence of science', *New Zealand Listener*, 4: 4.

Cornell, S. (2021) 'Foreword', in R. Joseph and R. Benton (eds) *Waking the Taniwha: Māori Governance in the 21st Century*, Wellington: Thomson Reuters, pp ix–xi.

Durie, M. (2005) 'Indigenous knowledge within a global knowledge system', *Higher Education Policy*, 18: 301–12.

Douglas, M. (2012) *Project Mātauranga*, [television series] Auckland: Scottie Productions.

Douglas, M. (2013) *Project Mātauranga*, [television series] Auckland: Scottie Productions.

Gerrard, J. and Kukutai, T. (2019) 'Foreword', *New Zealand Science Review*, 75(4): 61–2.

Hendy, S. and Wiles, S. (2021) 'An open response to "In defence of science"', *New Zealand Listener*, 23 July, Available from: https://archives.tane.harre.nz/masonbeenz/media/files/An%20open%20response%20to%20'In%20defence%20of%20science'%20New%20Zealand%20Listener.txt

Hikuroa, D. (2016) 'Mātauranga Māori – the ūkaipō of knowledge in New Zealand', *Journal of the Royal Society of New Zealand*, 47(1): 1–6. doi: 10.1080/03036758.2016.1252407

Hutchings, J. and Martin, W.-J. (2020). 'Foreword', *New Zealand Science Review*, 76(1–2): 1–2.

Jackson, A.-M., Hakopa, H., Phillips, C., Parr-Brownlie, L.C., Russell, P., Hulbe, C. et al (2019) 'Towards building an Indigenous science tertiary curriculum: part I', *New Zealand Science Review*, 75(4): 69–73.

Jackson, A.-M. and Mercier, O. (2020) 'Mātauranga and science II – introduction', *New Zealand Science Review*, 76(1–2): 3–5.

Jackson, S. (2013) 'Ko Te Houhanga a Rongo marae tōku tūrangawaewae: in search of a philosophical standing place for Indigenous development', Master of Arts dissertation, University of Otago, Dunedin.

Marsden, M. (2003) 'God, man and universe: a Māori view', in T.A.C. Royal (ed) *The Woven Universe: Selected Writings of Rev. Māori Marsden*, Otaki: Estate of Rev. Maori Marsden, pp 2–23.

McAllister, T.G., Naepi, S., Wilson, E., Hikuroa, D., and Walker, L.A. (2020) 'Under-represented and overlooked: Māori and Pasifika scientists in Aotearoa New Zealand's universities and crown-research institutes', *Journal of the Royal Society of New Zealand*, 52: 38–53. doi: 10.1080/03036758.2020.1796103

McKinley, E. (2008) 'From object to subject: hybrid identities of Indigenous women in science', *Cultural Studies of Science Education*, 3: 959–75.

Mercier, O.R. (2018) 'Mātauranga and science', *New Zealand Science Review*, 74(4): 83–90.

Mercier, O.R., Douglas, M., Rickard, D., Orlando, S., Morrison, S., and Apiata, D. (2014) 'Project Mātauranga – our science on-screen', in T. Black (ed) *Enhancing Mātauranga Māori and Global Indigenous Knowledge*, Wellington: NZQA, pp 69–86.

Mercier van Berkel, O.R. (2021) 'Transformations at the interface', in J. Ruru and L. W. Nikora (eds) *Ngā Kete Mātauranga: Māori Scholars at the Research Interface*, Dunedin: Otago University Press, pp 222–33.

Mika, C.T.H. (2012) 'Overcoming "being" in favour of knowledge: the fixing effect of "mātauranga"', *Educational Philosophy and Theory*, 44(10): 1080–92.

Ministry of Research Science and Technology (2007) 'Vision Matauranga: unlocking the innovation potential of Maori knowledge, resources and people', Wellington: Crown Copyright.

Morgan, K. and Manuel, R. (2020) 'Western science and Indigenous wisdom: is integration possible, practical, plausible?', *New Zealand Science Review*, 76(1–2): 6–12.

New Zealand Association of Scientists (2021) 'Mātauranga and science', Available from: https://scientists.org.nz/resources/Documents/PressReleases/NZAS-Ma%cc%84tauranga%20and%20Science.pdf

Ngata, T. (2021) 'In defence of colonial racism', Available from: https://tinangata.com/2021/07/25/defending-colonial-racism/

Pohatu, T. (2005) 'Āta: Growing Respectful Relationships', Manukau, Available from: https://ojs.aut.ac.nz/ata/article/download/121/101/

Rauika Māngai (2020) *A Guide to Vision Mātauranga: Lessons from Māori Voices in the New Zealand Science Sector*, Wellington, NZ: Te Rauika Māngai, Available from: https://www.maramatanga.co.nz/sites/default/files/Rauika%20Ma%CC%84ngai_A%20Guide%20to%20Vision%20Ma%CC%84tauranga_FINAL.pdf

Roberts, M. (1996) 'Indigenous knowledge and Western science: perspectives from the Pacific', paper presented at the Science and Technology, Education and Ethnicity: an Aotearoa/New Zealand perspective, Conference, Wellington, May 7–8 1996.

Simpson, L.R. (2004) 'Anticolonial strategies for the recovery and maintenance of Indigenous knowledge', *American Indian Quarterly*, 28(3–4): 373–84.

Smith, G.H. (1997) 'Development of Kaupapa Maori theory and praxis', PhD dissertation, University of Auckland, Auckland.

Smith, L.T. (1999) *Decolonizing Methodologies: Research and Indigenous Peoples*, London: Zed Books.

Stewart, G.T. (2019) 'Mātauranga and Pūtaiao: the question of "Māori science"', *New Zealand Science Review*, 75(4): 65–8.

Stewart, G.T. (2021a) *Māori Philosophy: Indigenous Thinking from Aotearoa*, London: Bloomsbury.

Stewart, G.T. (2021b) 'Defending science from what?', *Educational Philosophy and Theory*. doi: 10.1080/00131857.2021.1966415

Stewart, G.T. (2021c) 'Word weapons? Letters to editors', *ACCESS: Contemporary Issues in Education*, 41(1): 60–3. doi: https://doi.org/10.46786/ac21.3671

Tallon, J. (2020) 'Letter to editor', *New Zealand Science Review*, 76(3): 81–2.

Te Aho, F. (2021) 'Governance of Mātauranga Māori after Ko Aotearoa Tēnei', in R. Joseph and R. Benton (eds) *Waking the Taniwha: Māori Governance in the 21st Century*, Wellington: Thomson Reuters, pp 945–71.

Tohunga Suppression Act 1907, New Zealand Legislation.

Van Kerkhoff, L. and Pilbeam, V. (2017) 'Understanding socio-cultural dimensions of environmental decision-making: a knowledge governance approach', *Environmental Science and Policy*, 73: 29–37.

The Waitangi Tribunal (2011) 'Ko Aoteaora Tenei: a report into claims concerning New Zealand law and policy affecting Māori culture and identity', Wellington: Waitangi Tribunal, Available from https://forms.justice.govt.nz/search/Documents/WT/wt_DOC_68356054/KoAotearoaTeneiTT1W.pdf

PART III

The Decolonisation Agenda in Science Communication: Deconstructing Eurocentric Hegemony, Ideology, and Pseudo-historical Memory

9

Decolonising Initiatives in Action: From Theory to Practice at the Museum of Us

Brandie Macdonald and Micah Parzen

Introduction

The longstanding calls for museums to practise accountability, inclusion, restitution, and equitable access have grown into critical decolonial demands over the last two decades, and particularly since the 2012 publication of Amy Lonetree's book *Decolonizing Museums: Representing Native America in National and Tribal Museums*. Community members, practitioners, and scholars are collectively elevating the inherent need for museums to redress colonial harm and shift their operating practices from the colonial to the decolonial. However, many museums remain resistant to this much needed and imperative change. Why is that? Perhaps it's because change is difficult – it's uncomfortable, fluid, uncertain, and scary, especially if you have operated and been trained under the colonial constructs of 'industry best practices', which have reified your organisation's status as an 'authoritative expert' since its inception.

The 'museum' was established in order to support, legitimise, and celebrate the colonial endeavour. Museums grew from the European 'cabinets of curiosity', where the colonial elite would display their trophies of conquest in order to establish racial superiority and further justify the right to colonise. Much like today, in museums the objects, textiles, plants, animals, ancestral human remains, and animal skeletal remains were used to tell a story that the collector curated to match their interpretive plan, ethos, and worldviews (Bennett, 1995; Aldrich, 2009).

We recognise this barbarous past, acknowledge that colonialism continues to manifest in museums today, and thus ask ourselves the profound question: Can an inherently colonial museum be decolonial? Our answer is that we don't know, but we must try. Maya Angelou, American poet and civil rights activist, talks about how someone can only do the best they can until they know better, and that once they know better, they must also then do better. We agree with Angelou both on a personal and on an institutional level. We recognise that museums continue to actively perpetrate colonial harm towards Black, Indigenous, and Peoples of Colour (BIPOC) internationally. With a deep systemic colonial legacy, the Museum of Us, where we both work, is no different, and now it is our responsibility as senior museum practitioners (in key leadership positions), community members, and as humans to do better to redress this colonial harm.

In recognition of all of this, we hold on to the call for decolonial change and the need to do better as museum practitioners. In this chapter, we will be discussing how we work towards these efforts at the Museum of Us. We will begin by first locating ourselves through acknowledging the Indigenous lands that we reside/work on. We will follow by positioning ourselves and our ancestors in relationship to our connection to colonialism, our work, and our place in the world. We will then briefly speak to the history of the museum. We continue the discussion by talking about our work today and through providing tangible examples of how decolonial practices manifest within the museum's colonial walls and conclude with a brief overview of our commitment to the future.

Acknowledgement of land and self

We write this chapter on the unceded Indigenous homeland lands of the Kumeyaay peoples, over which they have stewarded for over a millennium. The Kumeyaay have suffered innumerable injustices from settlers, conquistadors, missionaries, by the US Federal Government, and by the hand of the museum. The land that the museum and the city of San Diego are located on is Kumeyaay land. Kumeyaay peoples were forcefully removed from this land in order for the museum to be built. We would be remiss if we did not state that we are all on Indigenous lands. Furthermore, we all benefit in various ways from the removal, dispossession, disenfranchisement, and active genocide of Indigenous peoples internationally.

We also recognise that colonialism, displacement, and genocide affect BIPOC in various ways, and the museum field (and in turn its practitioners and beneficiaries) have actively profited from this harm. We must be accountable to our ancestors, our place in the world (professional/personal), and how we are connected to colonialism. With regard to this complexity, we also hold space to locate and acknowledge ourselves by way of positionality recognition.

Brandie

I approach this work as an able-bodied, cis-gender, queer, Indigenous woman, who grew up in the foster-care/social services system as a ward of two states. I am an enrolled citizen of the Chickasaw Nation, with ancestral ties to both the Choctaw Nation and the Scottish Highlands. The Chickasaw Nation and the Choctaw Nation are federally recognised sovereign Indigenous nations residing within the settler colonial imperial borders of the United States.

Due to the imperial US legislation and colonial law for tribal federal recognition, in order to claim Indigeneity and be a citizen of a federally recognised tribe, I either have to prove my total percentage of 'Indian-blood' (also known as blood-quantum) and/or have proof of direct lineage descendant within federal and tribal documents. I have to hold two issued identification cards to prove my Native American Indigeneity, a 'Certificate of Degree of Indian Blood – CDIB card' and a 'Chickasaw Nation Tribal Enrollment Identification Card'. Furthermore, I am only allowed through the US legislation to be enrolled in one federally recognised tribe and legally claim it on certain documentation. These laws and regulations specifically target North American Indigeneity in an effort to 'breed out' North American Indigenous peoples and erase/delegitimise our existence.

I live and work in the complex colonial worlds of academia and museums. I hold a position of power and privilege professionally in the museum field, as the Senior Director of Decolonizing Initiatives at the Museum of Us. I am also enrolled in an Education Studies PhD programme at the University of California, San Diego, where I research the application of anticolonial and decolonial praxis in museums. I operate within and benefit from the structural colonial systems and the various manifestations of the colonial endeavour that I am concurrently working to disrupt.

I am a person who is grappling with the tension of being a transplant in a land that is not my own Indigenous homeland, and as an Indigenous woman navigating the very hostile, racist, and colonial fields of academia and museums. Additionally, I must be accountable to the privilege and power that I hold because of my skin's phenotype due to the presence of White supremacy. My lighter skin colour offers me the ability to blend into spaces and provides a protective shield of racial anonymity, making me more palatable to the colonial gaze. I recognise that many times this racial shield opens doors, spaces, and ears that are not accessible to my colleagues, friends, and relatives with darker skin. I would also be remiss if I did not recognise how this content is being written in the English language. This is a reinforcement of the dominant English linguistic hegemony that is a by-product of the colonial impact of cultural genocide within many of our ancestral communities (Thiong'o, 1992).

I am an educator, pracademic (a practitioner and academic), an aunty, and the next generation of elders. I am grounded in my Indigenous community's values of generosity, accountability, integrity, love, and perseverance. These values guide my path forward.

Micah

I am an able-bodied, cis-gendered White man of great privilege. I grew up in a secular Jewish household in La Jolla, an affluent community of San Diego, United States, raised by highly educated parents. I received a PhD in anthropology, then a law degree, and landed a job as an attorney at one of San Diego's most prestigious law firms, where I eventually became a Partner. I have been married to my wife for over 20 years, and we have two boys. By the time this chapter is published, I will be 52. For the past 11 years, I have served as the Chief Executive Officer of the iconic Museum of Us in Balboa Park. Doing so has been an extraordinarily humbling experience and the great honour of my life.

I've had my own hurdles to overcome in life, but (from the get-go) the deck was (and continues to be) stacked overwhelmingly in my favour. My multifaceted privilege paved a superhighway for my success. It afforded me countless opportunities denied to so many others, especially peoples of colour. I recognise that these opportunities came to me (and continue to come) on the backs of BIPOC, whose daily lives are oppressively burdened with inequities that I will never experience or fully understand. I know the benefits enjoyed by me (and my family – past, present, and future) simply by virtue of the privilege I was born into, have been at a shameful cost to so many other of my fellow human beings who do not possess those privileges.

Many years ago, I made a commitment to myself that, rather than denying that cost and continuing to be part of the problem, I wanted to learn how to be a part of the solution, if I could. My journey has been long, fraught with contradictions, and riddled with hypocrisy at its worst moments, and there have been plenty. I have struggled mightily and made many mistakes, but that struggle, and those mistakes have been my greatest teachers. I continue to grow into my understanding of how I often unwittingly wield my privilege in ways that are damaging to others. But I am also growing in my ability to listen to, reflect with, learn from, and ask peoples of colour (among others) whether I can be of help and, if so, how.

Colonial legacy: Museum of Us

Originally founded in 1915, the 'San Diego Museum of Man' (name changed to the Museum of Us in 2020) opened as part of the Panama-California Exposition, celebrating the opening of the Panama Canal. At the time, there were Kumeyaay peoples living in Balboa Park up to the early 1900s. The organisers of the 1915 Panama-California Exposition, however, deemed them insufficiently 'Indian' to suit the expectations of the anticipated throngs of tourists.

As such, Kumeyaay peoples were forcibly removed from their home in the soon-to-be Balboa Park and thus dispossessed from their territory and living on their land. The exposition organisers then shipped in Native American Indigenous peoples from various Pueblo tribes located in the south-western areas of the United States (Arizona and New Mexico) who fitted the expected 'brief' of what an 'Indian' was supposed to look and act like. An 'Indian Demonstration Village' was then set up so that the tourists engaging in the exposition experience could gawk at the Indians doing 'Indian things', which were defined by the organisers as dancing, making pottery, weaving baskets and rugs (Decolonizing Initiatives Department, 2019; SD History Center, nd).

The park's ubiquitous Spanish colonial architecture, which stands as a monument to the colonial endeavour, only reinforces this history of harm to BIPOC from which Balboa Park emerged – perpetuated even by the name itself, which is named after the Spanish conquistador Vasco Núñez de Balboa. In fact, the museum's very facility is a prime example of this architecture. The museum operates in the historic California Building and Tower.

The California Building is an intricate, ornate, and some would say beautiful example of Spanish colonial iconography. The building's façade has images of boats, flowers, and nine men – European colonisers – sculpted into its skin. Bertram Goodhue, the building's architect, specifically selected these nine men in an effort to uplift their legacies' impact on the colonial development and settler colonialism of the Pacific coast and Americas. The full impact of these nine colonisers lies not in the development of the Pacific coast but in the perpetuation of violent exploitation, genocide, abuse, and enslavement of Indigenous peoples throughout the Americas (Decolonizing Initiatives Department, 2019, 2020; Museum of Us, 2020).

The history of colonial representation and the replication of colonial harm did not stop at the façade of the museum. For decades, the museum refused to listen to Indigenous peoples' requests for repatriation, access, and restitution. A surface-level examination of the museum's accession records shows the evidence of over a century of legal donations and purchases. However, a closer examination of these transactions reveals how the museum supported colonial pathways for extraction – such as grave-robbing, exploitation, and the use of power and privilege for inequitable trade (just to name a few). Similarly, exhibits displayed ancestral bodies, established racial hierarchy under the guise of anthropology, and curated pictures of Indigenous peoples as static, frozen-in-time, and exotic. This history laid the foundation for how the museum's decolonisation process manifests today.

Our decolonisation work today

In 2011, following a change in leadership, the museum leased the exhibit 'Race: Are We So Different?' for a period of three months. This exhibit, developed by the American Anthropological Association in partnership with the Science Museum of Minnesota, looked at: (1) race as a cultural construct, rather than something that is biologically or genetically determined; (2) how, over time, that construct has been reified through our various institutions – academic, legal, financial, educational, and the like; and (3) the experience of race and racism in everyday life. The decision to bring this exhibit to the museum was transformational. Specifically, it was the first time the museum looked under the hood of its own institutional privilege. This set the foundation for our critically reflective work moving forward.

We assert that the action of being critically reflective is an essential practice in working towards being better and doing better. Critical reflection as a practice is more than looking at the history of the museum and the field at large. It is embarking on the ways in which the museum's collection, exhibition, and administrative practices have actively erased and silenced Indigenous voice, identity, and autonomy. Additionally, it is reflection that leads to restitution around how the museum's history (and the field) have supported White supremacy, racist/colonial legislation, and BIPOC's rights to self-determination (Garcia et al, 2019; Macdonald, 2022).

Today, our critically reflective decolonisation work focuses on five guiding principles: (1) truth-telling about and accountability for the harm we have caused; (2) repatriation and transfer of ownership to home communities; (3) policy and systems change; (4) representation of BIPOC at all levels of decision-making; and (5) reciprocity (Macdonald and Vetter, 2021; Macdonald, 2022). We are always looking for ways to apply these principles in new ways across the organisation, especially within our administrative section of the museum. Our colonial legacy is interdisciplinary and interdepartmental. As such, we firmly assert that our decolonisation work to redress our colonial legacy must be the same. It must transcend the work that happens in our education and cultural resources (collections) department; it should operate as an overarching framework across the scope and breadth of our entire organisation.

Decolonial work in museums is an inherently fluid, non-linear process. The nonlinear nature of decolonial work is present within the museum's guiding principles and how they physically manifest within the work, whereby the lines between each of these guiding principles are blurred and frequently overlap. In the following paragraphs, we provide four exemplars of how three of our guiding principles (policy and systems, truth-telling and accountability, and reciprocity) are specifically incorporated into our decolonisation work in the Museum of Us.

Membership model: 'Membership on Us'

Traditionally, museums have relied on membership programmes as a core component of their business and community engagement models – we are no different. Such programmes serve as so-called 'donor pipelines' and help segment audiences based on financial capacity, identifying visitors who are willing (and financially able) to support and drive the organisation's mission. At their core, membership programmes were intended to build a sense of elite community and exclusive belonging around an institution.

This structure inherently and intentionally creates a separation – the 'haves', the ones who receive the benefits of membership, and the 'have nots', the ones who do not. The more financial support a member provides, the more they benefit from membership programmes. This may include exclusive access to VIP exhibit openings, behind-the-scenes tours, or discounts at the store or with facilities rentals. We grappled with this separation as antithetical to our institutional values. In particular, it long felt in direct conflict with our decolonising initiatives, which seek to address the longstanding colonial inequities that the museum (and the field) has established, especially for BIPOC communities and other underrepresented marginalised groups (Parzen, 2020).

In June 2021, the museum launched a new programme, called Membership on Us, that did away with museum 'membership' as we know it by providing unlimited general admission for one year to anyone who purchases a ticket. In this way, Membership on Us exponentially increases access to our museum offerings and our value proposition and simultaneously centres inclusivity and builds a system for reciprocity. As part of the Membership on Us programme, partnerships with underserved schools that typically do not access museums allow their students, faculty, and staff unlimited and free access. Another important component of Membership on Us is unlimited free access for all Indigenous peoples. Finally, we have a 'pay what you wish' offering for anyone experiencing financial hardship (Parzen, 2020; Kragen, 2021; Membership, 2021). Membership on Us is a representation of both our commitment to equity and decolonial action in the museum field.

Decolonisation Initiatives in Action signage

In taking time to critically reflect, we asked ourselves several questions:

1. Are we being transparent with our external communications around our internal decolonisation work on exhibitions?
2. What would an ethical transparent practice be for external communications?
3. How do we ensure that when we talk about our decolonisation practices, we communicate an accountability-based narrative that it is the responsibility of the museum, due to its colonial legacy?

a. Particularly, one that would be honest about our dated colonial exhibits that have caused harm, and create an entry point for decolonisation work to visitors, without being a performative 'pat on the back' that overshadows the coloniality of the museum?

What this reflection led to was the development of our Decolonising Initiatives in Action signs. These signs highlight and illustrate the need for accountability and truth-telling about the colonial legacy of the content of exhibits and help to engender educational access points through curation for the general public (Macdonald, 2022). For example, the museum team is currently working with Maya community consultants to reframe and redefine an archaic permanent exhibit titled 'Maya: Heart of the Sky, Heart of the Earth'. This exhibit portrayed the Maya peoples as people of the past, without any contemporary contextualisation of them as a living and thriving diverse community that has been dispersed all over the world due to colonialism.

There are three sections to the Decolonising Initiatives in Action signs that integrate the first decolonising initiatives guiding principle. These areas are: (1) truth-telling about the colonial legacy; (2) transparency around the funding source; and (3) accountability through commitment and consultation. The truth-telling aspect of the signs talks to the colonial aspects of the exhibit and how it replicates harm. For example, the sign directly points out that the exhibit is outdated, how it is from the viewpoint of a Euro-American researcher and not that of the Maya peoples, and how Maya people are still here and, with their communities in the diaspora, are actively practising their cultural traditions in all areas of the world.

The second section of the decolonisation signs talks about who is funding this specific project. This practice is not included simply because we wish or need to elevate the sponsor for promotional purposes. We include the funder's name in these signs because it is important for our decolonisation work to be transparent around where our funding sources are coming from and how they might also be connected to (support) the colonial endeavour.

The third section of this decolonisation signage is the declaration of our commitment and a direct reflection of how we approach this work. The commitment section is directly copied from the joint agreement that was collectively edited and approved by our Maya consultants. For example, the commitments presented on the signage speak to how the museum will prioritise the Maya community voices, realities, and requests when displaying, interpreting, and writing content for the revised exhibit; and how the museum will recognise and honour Maya sovereignty and knowledge (oral and written) as the correct and most legitimate source of data when defining the importance of any cultural heritage and interpretive planning; and that Maya consent is at the forefront of all decision-making.

Holiday leave: disrupting Eurocentrism and White supremacy

The museum has provided 13 paid holidays each calendar year, which have in turn been the designated days that the institution has mandated its staff members to collectively take time from work. However, through our decolonising initiatives work, engaging in the collective practice of critical reflection, and through the guidance of many of our BIPOC staff, the museum has come to understand just how many of those mandated holidays are deeply rooted in colonialism and White supremacy. For example, New Year's Day, Presidents' Day, US Independence Day, Thanksgiving, Christmas, and so on. In response, we realised that we must do better for our team and revise our holiday policy to be more representative of all peoples and thus disrupt the systemic social practices that legitimise colonialism.

On 1 January 2022, the museum launched a new holiday policy called Holidays for All, which will create flexibility for members of staff to substitute (and even group together) holidays that are significant to them and their own cultural/religious/racial groups. Team members may still choose to take any or all of the historically designated holidays, if they wish. However, they will also have the autonomy to decide what their leave for the year looks like and how it aligns with their ontological and epistemological needs. We are delighted at the opportunity to initiate this innovative transformation and thus change the embedded colonial practices within our human resources department so that it better supports all of our staff members in a more decolonised way (Parzen, 2020). This policy, along with our new membership model (Membership on Us) and the following community-centred leave policy, are all exemplars of our third decolonising initiatives guiding principle, which focuses on the need to change colonial institutional policies and systems.

Community-Centred Leave

The decolonisation conversations around human resources colonial practices inherent in the museum have not stopped at solely revising our holiday leave. Like most US employers, the Museum of Us offers various types of paid leave, which includes vacation and sickness time. However, when sitting at the leadership table with our Indigenous staff members, they shared how they needed time off to participate in multi-day ceremonies in their home communities. The administration's initial reaction was to direct them to the museum's vacation and other leave policies. Our Indigenous colleagues pointed out the obvious in these meetings – which was that their absence from the office was neither 'vacation' nor 'sickness' time. On the contrary, their time in the community was/is hard work and is a deep commitment that is centred on cultivating and sustaining community well-being – the well-being of the past, present, and future generations. It was for the greater good and not solely focused on their personal enjoyment or fulfilment.

We soon realised that this need and community-centred practice is not isolated to just our Indigenous colleagues, and that if we are going to be an organisation that centres decolonisation practices, then we must also review how Eurocentrism is imbedded in all areas of our leave practices, and not just holidays. We considered classifying this leave in a variety of ways and worked diligently to ensure that our decolonisation policies were also in alignment with (at times, unfortunately, colonial) employment laws. We ended up developing a new paid leave category altogether. We call it Community-Centred Leave, and it is specifically designed to encourage *all* employees to support their communities in ways that are meaningful and culturally significant to them. This decolonial policy was effected from 1 January 2022, whereby every staff member at the museum will be able to take up to three days of paid Community-Centred Leave per year (Parzen, 2020).

Conclusion

It is all too easy to be overwhelmed and feel defeated by the magnitude of how deeply colonialism is embedded in the museum field, and within the organisations we work at daily. We hear you, we see you, and we can absolutely relate. Concurrently, we find strength in solidarity and in knowing that with each step forwards (albeit, at times, incredibly challenging), we are taking transformative steps towards building a better present and future.

The Museum of Us is a 105-year-old cultural anthropology museum with a deep colonial legacy that has harmed so many BIPOC peoples since its inception. We have made the commitment towards redressing this colonial harm through our decolonising practices. We are ten years into this journey of becoming a more decolonised organisation and have the next five years of our decolonisation work outlined in our Decolonising Initiatives Strategic Action Plan. Our practices started as a theoretical effort to move the museum towards decolonisation as an endpoint but have transformed into a paradigm shift that recognises this decolonisation work as being more than just a grant deliverable, it is our work moving forward for the next 100 years.

We assert that decolonisation is a collective endeavour, at its core. Additionally, we see decolonisation as a verb. It is actionable accountability and truth-telling. The decolonisation of museums requires a deep and unyielding commitment (financially, structurally, and culturally) to continuous action in a forward-moving process of reckoning with the ever-present traumatic legacy of colonialism.

References

Aldrich, R. (2009) 'Colonial museums in a postcolonial Europe', *African and Black Diaspora: An International Journal*, 2(2): 137–56. doi: https://doi.org/10.1080/17528630902981118

Bennett, T. (1995) *The Birth of the Museum: History, Theory, Politics* (Kindle edn), New York: Routledge.

Decolonizing Initiatives Department (2019) 'Decolonizing initiatives, Museum of Man', San Diego Museum of Man/Museum of Us, Available from: https://www.museumofman.org/decolonizing-initiatives/

Decolonizing Initiatives Department (2020) 'Colonial legacy: the museum's facade', Google Arts and Culture Exhibit, 17 December, Available from: https://artsandculture.google.com/exhibit/colonial-legacy-the-museum-s-facade/cgJC4cUKV7q4Kw

Garcia, B., Hyberger, K., Macdonald, B., and Roessel, J. (2019) 'Ceding authority and seeding trust', American Alliance of Museums, Available from: https://www.aam-us.org/2019/07/01/ceding-authority-and-seeding-trust/

Kragen, P. (2021) 'When Museum of Us reopens Wednesday, it will unveil more than a new name', *San Diego Union-Tribune*, 18 April, Available from: https://www.sandiegouniontribune.com/communities/san-diego/story/2021-04-18/when-museum-of-us-reopens-april-21-it-will-unveil-more-than-a-new-name

Macdonald, B. (2022) 'Pausing, reflection, and action: decolonizing museum practices', *Journal of Museum Education*, Available from: https://eric.ed.gov/?id=EJ1338177

Macdonald, B. and Vetter, K. (2021) 'From the colonial to the decolonial: the complex intersection of museum policy and practice', *South African Museums Association Bulletin*, Available from: https://eric.ed.gov/?id=EJ1338177

Membership, Museum of Us (2021) 'Membership at the Museum of Us', Available from: https://museumofus.org/membership/

Museum of Us (2020) 'Museum of Us California Building', Museum of Us, About Us, Available from: https://museumofus.org/history#our-home-the-california-quadrangle

Parzen, M. (2020) 'Knowing better, doing better: the San Diego Museum of Man takes a holistic approach to decolonization', American Alliance of Museums, 8 January, Available from: https://www.aam-us.org/2020/01/08/knowing-better-doing-better-the-san-diego-museum-of-man-takes-a-holistic-approach-to-decolonization/

SD History Center (nd) 'Panama-California Exposition', San Diego History Center, About, Available from: https://sandiegohistory.org/archives/amero/1915expo/ch5/

Thiong'o, N. wa (1992) *Decolonising the Mind*, Nairobi: East African Educational Publishers.

10

Falling from Normalcy? Decolonisation of Museums, Science Centres, and Science Communication

Mohamed Belhorma

Introduction

A social construct is always created at the centre of any communication framework. Science communication is no different in this regard, despite its goals and the evidence-based scientific facts it presents. Like any other kind of communication, science communicators inevitably create a frame and pursue an agenda. The creation of a frame falls within the realm of normative action and normativity. In our dictionaries such as Merriam-Webster we can read that normal is defined as conforming to a pre-existing standard, 'a type, standard, or regular pattern: characterised by that which is considered usual, typical, or routine' (Merriam-Webster, 2021). The general understanding is that the norm is the usual or common, while normal is opposed to abnormal. In contrast to these two terms, normative refers to a morally approved ideal, or 'what ought to be' from an ethical point of view.

From a philosophical perspective, normative ethics analyse how people should act, while descriptive ethics analyse what people think is right. The problems of science communication do not lie in descriptive ethics: there is a clear consensus on the ethical problem of Eurocentric hegemony and its consequences. Science communicators are well aware of normative aspects as long as they are related to the disciplines they communicate about. These understandings beg the following questions: what role does our own normative frame play when it comes to decolonising our field? If it does have a role, then why is it poorly addressed? Does science communication have a blind spot when it comes to its own normality, norms, and normativity? This chapter seeks to explore some possible reasons for this blind spot in science communication in relation to the normative ethics of decolonisation. If these epistemic norms are grounded in the colonial purpose of domination and hegemony, demonstrating and elaborating some of their impact on non-Western populations could help identify the problematic influence of unchallenged norms.

Postcolonial literature has defined the causes of this historically unchallenged state of normativity in the Western world. In this chapter, practical case studies from the scholarly communication of museums, and other areas of communication such as popular culture within the same framing analysis, will illustrate how systematic and widespread the problem is. The work of Gloria Wekker (2016) will be used to demonstrate how unquestioned colonial norms have created a sense of innocence that limits the possibilities of challenging said norms. The subtitle of Edward Said's (1978) *Orientalism*, 'Western conceptions of the Orient', epitomises that instead of epistemology we have an artificial construct that has helped justify colonialism. By combining the concepts of innocence (Wekker, 2016) and the cultural archive (Said, 1993), the chapter will explore how science communication is able to live comfortably with its blind spot on normativity.

Finally, this chapter assesses the lack of epistemic diversity in science communication and the hegemonic/exclusive position of Western epistemic and Western epistemology, regardless of the quality or relevance of other knowledge production. Whether it is communicating epistemology, the erasure of non-Western knowledge, or the interpretation of theories, there is a whole field of reflection that science communication must explore in order to deliver transformative decolonisation.

Overview and situational problem analysis

The Museum of European Normality (2008) is an immersive art installation by artists Maria Thereza Alves and Jimmie Durham.[1] The work offers the visitor a museum experience that reverses the roles of exhibition object and viewer. The immersive installation enables the interrogation of exhibition methods by making visible the unequal power relations associated with the act of exhibiting itself. The work shows us that decolonisation is not limited to restitution or attribution. It highlights the fact that colonial violence can be contained in the way science is exhibited and communicated. What Europeans perceive as neutral, positive, or even aesthetic is perceived by others as a brutal act of violence.

This art installation is another reminder of colonial double interactions – the dominant and the subaltern (Spivak, 1988). In the dynamic processes inherent to this dual scenario, and within the contexts of postcolonial transitioning, providing other cultures with their rightful space and giving back a voice to silenced civilisations are practices that do not erase the pre-existing narrative. It merely creates an alternative narrative that has to coexist alongside the dominant one. Furthermore, these strategies only indirectly question the dominant side. W.E.B. Du Bois (Du Bois, 1910) made this observation at the beginning of the previous century and thus opened up a wide field of reflection by integrating both sides of the racial divide. He argued in favour of providing visibility and space for the subaltern but also for equitable access in order to interrogate and challenge what perpetuates and enforces privileges.

Colonisation was an act of exploitation, dispossession, and systematic destruction. To justify the colonial endeavour, the Western world had to create an ideological framework that rationalised the exploitation of the rest of the planet. To this end, everything in the realm of culture and knowledge was rewired to justify colonisation. Not only did the West create a different narrative concerning the identity of the colonised but, as Said (1978) elaborated in *Orientalism*, it even tampered with basic civilisational features such as time itself. Johannes Fabian (Fabian, 1983) shows how anthropology was used to justify colonialism by creating a different temporality and thus to deny a simple scientific fact: coevolution within a similar time. To justify colonisation, science claimed that people around the world did not live in the same temporality.

Cultural imperialism has led to alienation on one side (that of the subalterns) and blindness to the dominant normativity on the other, and this can only be overcome by examining normativity and its challenges. Since the beginning of the 21st century, postcolonial academic reflections have increased in number and quality. Concepts like that of innocence (Wekker, 2016) and cultural archives (Said, 1993) show how convenient and easy it is to consider the dominant frame as the normal frame. However, decolonising science communication is not as simple as removing the colonial frame from the epistemological frame, given that it is often difficult to recover the alternative epistemology associated with the 'colonial-era' artefact being shown.

For over five centuries, the pervasive and systemic process of domination and oppression that has reshaped the face of the planet by erasing civilisations while reshaping other cultures and knowledge for the benefit of a few has provided museums with more challenges to address than the mere inappropriate presence of certain objects in their institutions. Science museums that own a collection cannot escape the issue of colonialism. Every museum in the Western world that is old enough owns or has owned looted artefacts and/or human remains. Meanwhile, institutions that do not own any collections very often fail to see how decolonisation relates to them. This is why, when this aspect is brought into discussions when we create content in science museums and centres one can hear from colleagues reactions such as, there is no need to bring the subject of colonialism into an exhibition about Einstein or Marie Curie or that an exhibition about Space or atoms has nothing that needs to be decolonised.

This failure to see and acknowledge the need to decolonise science museums and centres is also at the root of the problem. Science museums and centres without collections should learn from performative museology and ethnographic museums that have developed strategies that enable them to undertake decolonisation. The collaborative approach and other inclusive strategies developed in ethnographic institutions that help move the process forward are crucial steps. This transformative approach (Price, 2009) gives curatorial leadership to the community and provides ethnographic museums with strategies to launch the long decolonisation process. Performative museology attempts to decolonise by either exposing how museum institutions treat the subaltern or by asking subalterns to lead the narrative on their collections. Yet, very few of these good practice strategies are applied by science communication in general and science museums and centres without artefacts in particular. Science museums, centres, and science communication lack a catalogue of equivalent strategies in order to be as productive in decolonisation as ethnological museums or museums with collections. As a result, science museums and centres and science communication are falling behind on this critical front.

Normalcy and falling from normalcy: definitions, concepts, and history

Normalcy and cultural imperialism

In order to gain awareness of the need to decolonise in the context of the production of science communication, it is useful to look within the framework of privileges and constructed supremacy. Du Bois (1910), Baldwin (1998), and Roediger (1999) provide analyses that enable a closer examination of what race theorists call ‘unjust enrichment’. This process facilitates the exploration of why ‘unjust enrichment’ biases are not questioned, remain embedded in dominant knowledge systems, and continue to be passed on to the next generation. Within colonialism, there is a particular aspect that is labelled cultural imperialism. Cultural imperialism is a type of dominance that extends beyond military force or economic exploitation. It entails aspects of latent dominance through subliminal frameworks such as educational channels and media formats. These formats and channels have subtly but constantly and powerfully conveyed the superiority of the coloniser’s culture and science. For centuries, the constant application of cultural imperialism created a popular culture based on biased and limited knowledge.

Over time, this limited version of popular culture has achieved the dominant status of 'normalcy' and a new truth that was no longer debated, interrogated, or challenged. The profound legacy of colonialism is deeply ingrained in educational curricula and popular culture across the globe. These curricula have been taken for granted to such an extent that their legitimacy and accuracy are hardly questioned. Said (1993) used the lens of literature in *Culture and Imperialism* to describe in detail how culture is foundational to imperialism. Literature, and, by extension, culture, have 'the power to narrate, or to block other narratives from forming and emerging', with the consequence that, 'in turn, imperialism acquires a kind of coherence, a set of experiences, and a presence of a ruler and ruled alike within the culture' (Said, 1993).

Imperialism becomes an invisible frame of reference that shapes the culture of the entire empire, coloniser and colonised alike. Said (1993) listed many French and Anglo-Saxon classical literary fiction and novels embedded with colonial ideas and tropes. He found that they can be present in the most significant works of literature and yet remain unnoticed, presumed harmless while they keep flowing within popular culture.

Reinterpreting museum collections and/or restitution is obviously no longer enough. The museum community is actively acknowledging the impact of cultural imperialism beyond the collection case. A global shift in both attitude and understanding of the colonial past has compelled museums to re-examine their contribution to that particular narrative. In 2019, the International Congress of Museums (ICOM) introduced a new and transformative definition of the museum's role in this regard.[2] This conceptualisation fuelled a heated debate on how museums define themselves. The ICOM proposal had the invaluable merit of addressing the 'elephant in the room' in two ways. First, it exhorted the ICOM community to address the colonial past and the subsequent role of the museum at the semantic level. The second merit of the proposal was that it attempted to adapt this new definition to a longstanding structural and systemic issue, within a 'mainstreaming' approach.

Normalcy as a multiplier of 'business as usual' inertia

Like any other systemic issue, addressing colonialism and its remnants requires comprehensive tools and measures. Simply put, the best solutions are, for the most part, at a level that is difficult to address. Museum collection restitution is the most iconic example of this situation and is the most discussed topic as well. Institutions and governments work actively in setting protocols to initiate the restitution process. Yet, in many instances these efforts come attached with a catch-22 of the collection's ownership by the state and not the institution. This argument then often factors in questions as to the nature of future conservation conditions in the country of origin – in some cases, using these potential challenges to justify delays to restitution efforts.

In addition, when an issue is systemic, it tends to limit the impact of local events or actions. Institutions have to consider the benefits gained from investing energy and resources in work that – due to systemic externalities – will make a relatively small contribution to their key performance indicators compared to other projects. Science centres, museums, and science communicators need to justify their actions and show results. Simple managerial decisions that favour the best strategy to increase the number of visitors and their satisfaction levels create a dilemma between investing for the captive audience, for several different niche audiences, or for a wholly diversified audience profile.

It is important to realise that these issues question the very identity of the museum by engendering transformation in its juridical framework and its collections. This is what the new definition of the museum submitted to the ICOM was about – a paradigm shift. This transformative definition failed to garner the required majority votes of ICOM members in 2019, and subsequently, a revised 'compromise' version was adopted in 2022, by ICOM.[3] When it comes to questioning the very identity of the museum as encyclopaedic and universal, science communication, science museums, and science centres are very much concerned parties.

On not being the norm

Frantz Fanon (1963) and Albert Memmi (1965) describe the colonial situation from a socio-psychological perspective. They both describe the colonised and the coloniser as opposing poles. 'Colonization has produced the colonized just as we have seen it produce the colonizer' (Memmi, 1965). They also agree that the psycho-sociology of the coloniser and the colonised is intertwined with the causes and consequences of colonisation.

In order to discuss normalcy and what it entails, it is first crucial to understand what being estranged from this normalcy implies. Ngũgĩ wa Thiongo's (2011) description of this alienation in the cultural context is profoundly insightful: it is the isolation from any positive learning, not being a model, and very often being the embodiment of everything that opposes normalcy. Every aspect of child development and interaction with culture and education is thus defined by alienation. Both Fanon (1952, 1963) and James Baldwin (1998, 2010) describe in detail this damaging legacy of disconnection, disenchantment, and alienation inherited through the colonial experience.

It is important to bear in mind that around 90 per cent (Sarr and Savoy, 2018) of the cultural heritage of Africans is currently located outside of their continent. Their access to this heritage is thus denied. Yet, alienation does not only stem from a lack of access to this heritage. It also depends on how this heritage is displayed, elaborated, and communicated. Most importantly, alienation lies in the storytelling and how this narrative is imposed and informed wholly from the perspective of the dominant coloniser. Thus, cultural imperialism normalises facts, beliefs, and versions of the truth in the dominant culture and 'colonises the minds unnoticed' (Barthes, 1957).

On enjoying colonial normativity

Museums 'speak' from a position of authority. Akin to fictional literature, they are content purveyors. Despite a dislike for hierarchy, museums have a top-down relationship with their visitors. Museums are created as temples (Pomian, 2020, 2021) built to be purveyors of truth and guardians of verified and orthodox history. Consequently, curators assume the role of 'authority' (Unsal, 2019). The narrative that curators establish under a label of scientific objectivity and expertise holds a different weight. It also affects the public's understanding of science. In their studies, Bourdieu and Darbel (Bourdieu et al, 1966) revealed that even people who never visit museums credit the institution with expertise in the field they display. Acknowledging the impact of authority and expertise in the legitimisation and perpetuation of colonial biases is essential to understanding the alienating effect of the dominant culture.

In *White Innocence: Paradoxes of Colonialism and Race*, Gloria Wekker (2016) explores normativity using Said's concept of cultural archives:

> [W]hat Said is referring to here is that a racial grammar, a deep structure of inequality in thought, was installed in the nineteenth-century European imperial populations and that it is from this deep reservoir, the cultural archives, that, among other things, a sense of self has been formed and fabricated. With the title White innocence, I am invoking an important and satisfying way of being in the world. (p 2)

According to Wekker, centuries of imperial rule built up an unacknowledged reservoir of knowledge based on 'dominant meaning-making'. This normativity as described by Wekker has two fundamental features. First, it is unacknowledged, taken for granted like a convention and hence never discussed, questioned, or debated. Secondly, it is a satisfying and comfortable way of being in the world. In this state of normalcy, there is no discomfort. It is built to make the dominant comfortable. It allows the dominant to live with an arsenal of justification and rationalisation of structural inequality at its disposal that enables feelings such as innocence, confidence, and dissociation. In *Standard White: Dismantling White Normativity*, Michael Morris (2016) showed that this taxonomical aspect of normativity creates a disparity in access to care between those who belong and those who do not belong to that norm – making the healthcare system potentially life-threatening for people outside of the standard White norm. On the contrary, if one belongs to the norms, one can enjoy the highest standards of care.

Cultural imperialism – and by extension being privileged – has created a set of normativity with such positive attributes that the culture is no longer just the norm; it has become the ideal (Mills, 1997). This 'quality of the norm' influences how anything that falls outside of normalcy is perceived and received. Wekker (2016) refers to the case of 'Zwarte Piet' (Black Peter) in the Netherlands and the extraordinary reactions to the demands of People of Colour that Dutch society desist from normalising this longstanding tradition/celebration given what they see as its racist overtones and connotations. She claims that questioning the consensus over the Dutch description of 'Zwarte Piet as harmless and innocent' was complicated and risky, even when such dialogue was exchanged with other academics within a paradigm of peer-reviewed and collegiate discourses. This shows that normativity created by colonialism and cultural imperialism cannot be discussed like other taxonomies where the organisation of the norm is based on objective criteria. There are objective and quantifiable losses to not being in the norm, namely the loss of attributes, value, recognition, respect, validation, and affirmation.

Case studies: normativity of cultural imperialism

Goodwill and the 'blindness' of Eurocentric privilege

'Zwarte Piet' in the Netherlands, as highlighted in the previous section, is not the only display of offensive, degrading, and unscientific storytelling that normativity has made acceptable and even positive. A multitude of other examples exist that are specific to museums and culture. One such instance is the first attempt at decolonisation by the Royal Ontario Museum (ROM) in Canada.[4]

In 1989, ROM created an exhibition entitled 'In the Heart of Africa'. The museum's intention was to use this 'unique' exhibition to challenge the White settlers' perspectives and narratives of the Canadian scenario. In this regard, the curators used methodologies and approaches such as irony and provocative texts to challenge the traditional and stereotypical views of Africa and people of African descent held by the Canadian population.

Unfortunately, their good intentions were not enough, and they inadvertently created the most iconic exemplar of 'what-not-to-do' in an exhibition that seeks to challenge the colonial narrative (Schildkraut, 1991). They ended up in a scenario whereby they had managed to offend diverse factions from every part of the sociopolitical and cultural spectrum. Public protests ensued, some of them violent, and every museum that had booked to host the exhibition cancelled. This case became the epitome of the consequences of working within unchallenged normativity. The curation team had been genuinely convinced that they were challenging the imperialist vision and narratives with wit, bravado, and irony.

For more than 27 years, ROM and the exhibition makers – like the director of ROM, John McNeill – repeatedly claimed in their defence that the In the Heart of Africa exhibit was 'clearly and very consciously anti-racist' (Vincent, 1990). Their failure to understand or acknowledge their shortcomings clearly stemmed from an inability to recognise that oppression and cultural imperialism were operating within their zeitgeist, norms, and institutional frameworks. The world had to wait until 2016 for the institution to apologise.[5] ROM acknowledged the problem 30 years later when societal norms had shifted towards a more transformative understanding of the colonial legacy through the efforts, in large measure, of civil society organisations. The ROM case, then, is not an exemplar of an active institutional (internally driven) fall from normalcy but rather that of an institution that was driven to shift from normalcy by external pressures and societal censure.

"'Black movements are censoring art"': the case of Exhibit B

Wekker (2016) reflected on two attitudes that are characteristic of the problematic nature of the dominant norms. The first is the ignorance of the dominant norms, and the second is the acknowledgement of the dominant norm but a failure to examine the cultural archive on which it is built and sustained. As a result, we are unable to critically reflect on the issues. Consequently, the reactions of anyone who challenges said norms are deemed emotional or subjective. The most illustrative cases to date involve the White South African artist Brett Bailey,[6] who designed an installation titled Exhibit B, as a continuation of his first work, Exhibit A, and the ROM exhibition titled 'In the Heart of Africa'.

Bailey's first exhibition was meant to serve as a critique of human zoos. Exhibit B similarly exhibited a series of stills of Black people while claiming that its aim was to challenge the display of Black individuals in Museums throughout history.

In 2014, the Barbican Centre in London decided to cancel its hosting of Exhibit B due to the controversies and protests against the exhibition, enacted mainly by Black British communities. Exhibit B also sparked a similar reaction from Black French communities when it was displayed in Saint-Denis, France.

Both the Barbican Centre in the United Kingdom and the Théâtre Gérard Philipe in France condemned the protestors and activists. They described the protests around Exhibit B as an act of censorship and an attempt to hinder freedom of expression. Yet, notwithstanding these statements, the Barbican Centre still decided to cancel the event in London.

The professionals behind the In the Heart of Africa exhibition in Canada and Exhibit B in Europe believed that their projects were attacking colonialism and racial oppression. Yet, when they were confronted with protests from Black communities (the very peoples that they claimed to be championing with their exhibits) and with negative reactions from the masses, the defensive and self-justifying status quo of dominant culture remained.

To this day, many in the arts and the theatre scene in France view the Exhibit B episode as an example of censorship and 'cancel-culture'. Very few professionals in the field have objectively discussed the points and issues inherent in Exhibit B that the Black communities in France and the United Kingdom sought to challenge. It is interesting to compare the two works – the reflection of the Museum of European Normality on the very grammar of exhibiting as the problematic factor and purveyor of violence. Meanwhile, Exhibit B kept the grammar of exhibiting unnoticed and put the problematic factors on the viewer's gaze regardless of who the viewer might be. In seeking to understand the underlying perspectives behind the protests with regard to Exhibit B, critical questions for consideration arise such as: can a White South African artist use Black people to illustrate oppression that shaped *his* normalcy in a different way than the people he exhibited? Is access to art museums and curators for other Black/African artists equivalent to that of Bailey's? Were the resources needed to work on such a project made equally available to other artists? According to Françoise Vergès (see Maxime Cervulle, 2017), the curation of Exhibit B is an example of dispossession and cultural expropriation rather than decolonial content.

Popular culture and museums: *Indiana Jones*, *Black Panther*, and what pop culture reveals about normalcy

Indiana Jones, archaeology, and ancient civilisations

Popular culture also offers interesting and illuminating examples that shed light on museums and the work of their professionals with regard to dominant culture and its normalcy.

Much can be said about the *Indiana Jones*[7] films in terms of casual colonialism as a genre and as a character. These characteristics are on display in terms of how the *Indiana Jones* films illustrate colonialism in science communication, and more specifically, in how the character is used to interpret archaeology. Archaeologists use *Indiana Jones* to illustrate what their job is not.[8] Articles, exhibitions, and documentaries show that the Indiana Jones character does not behave like a scientist. The character visits archaeological sites all around the world and is confronted with state-of-the-art science that non-Western civilisations have developed. This aspect of the *Indiana Jones* narrative is one that seems routine, and yet it is not. This might not seem like an issue worth exploring, until what lies at the end of the spectrum of this relativism gets in the way of science communication. A classic example here is the case of the theories displayed in the 'Ancient Alien' shows. Theories from this show strip pre-colonial African civilisations of their scientific heritage and attribute them to aliens, instead.[9] The global history of discrediting science myths around colonised civilisations could benefit from the same demystification work that was done around the profession of the character of Indiana Jones. Yet, it remains a very rare theme in museum institutions, and these issues are seldom addressed within the science communication profession.

'Black Panther', *museum artefacts, and the 'cup of coffee'*

The *Black Panther* movie is a ground-breaking Hollywood film, and one that has generated many reactions, commendations, and content re-production among decolonial activists. The film features what some have deemed to be a particularly controversial scene that takes place in a museum – the action in the scene involves the museum heist of a looted African artefact – and has brought to the fore and into the realm of popular 'mainstream' discourses the issue of the restitution of looted artefacts in Western museums.

This scene engendered much discourse and impact on filmgoers and also sparked a deep dialogue among museum curators. The heist of the looted African artefact and its legitimacy, legality, and ethics raised many moral and legal questions around colonialism.

Museum professionals also debated and spread hashtags on social media regarding another scene in the film – the behaviour of the curator who brought a cup of coffee to the galleries[10] was deemed shocking to museum professionals. Bringing food and beverages close to an artefact that a curator is commissioned to preserve is one of the most unprofessional actions that a curator can undertake.

Museum professionals who spread hashtags and trends around the coffee cup illustrated to the world a highly indicative exemplar of museum normativity – that is, the triviality of the presence of looted artefacts in a gallery, while a cup of coffee inside the same gallery is deemed a much more vexatious and abnormal presence.

Coming a day after the World's fair, why do inaccurate narratives remain the norm?

The award-winning Hollywood film *Hidden Figures*, released in 2016, features the untold story of three African American women in science, technology, engineering, and mathematics (STEM), who worked as pioneers and innovators of the US-NASA space programmes in the 1960s and 1970s, particularly the NASA moon landings. This insightful and illuminating film has contributed immensely to raising awareness among science communicators of the imperative to communicate the women in science narratives through the more inclusive equity lenses of the intersections of gender *and* race. Through this landmark book and film, the names of pioneering Black women in STEM, such as Mary Jackson, Katherine Johnson, and Dorothy Vaughan, are now as familiar as those of their White counterparts. The question for science communication is why they were not as well known as their counterparts for almost half a century.

The historical analysis and framing of normalcy lay a crucial foundation to the cultural imperialistic aspects of science communication. This includes the unacknowledged lack of content, reference, or illustration that comes from outside the Western frame of reference with or without collections and regardless of the topic.

Foucault asserts that '[t]he history of some is not the history of others' (Foucault, 1976), and in his view, counter-history is aimed at breaking the continuity of glory and is composed of a principle of heterogeneity, accompanied by a principle of discontinuity (Foucault, 1976). The 'continuity of glory' is at the core of what the French describe as a 'mission civilisatrice' (civilising mission). It is a subjective and biased version of history with the sole aim of justifying colonialism and its consequences. This narrative starts with the postulation that the West is the cradle of civilisation. The West bestowed immense favours upon the rest of the world by sharing its civilisation with less fortunate populations. This version of history has shaped every aspect of dominant culture from the perception of the West, in relation to the peoples and cultures of the colonised worlds.

The perpetuation of the 'continuity of glory' ethos means that institutions keep using the same set of cultural archives.

The Science Museum Group, in the United Kingdom, committed very early on to a reflection on equality and inclusion. Their work has had a positive impact in a number of areas with regard to diversity and inclusion in science communication and science centres and museums. In 2012–13, they showcased the exhibition 'Codebreaker: Alan Turing's Life and Legacy' with an in-depth historical contextualisation of Alan Turing's life that included his sexual orientation and how it affected the way he was treated by the scientific community and society at large given the profound level of homophobia prevalent in society during his lifetime.

In terms of gender inclusion, the names of women scientists and pioneers, such as Ada Lovelace, Mary Somerville, Florence Nightingale, and Rosalind Franklin, are now frequently heard of in science museums and science centre galleries. Yet, this counter-history has taken many decades to move to the forefront and to permeate the science museum and science centre worlds. The frame-shifting on gender equality in science has only occurred long after historians revised the role of women in science and technology and long after activists have brought this knowledge to the public space. Learning about scientists' sexual orientation or hearing the name of a female scientist aside from Marie Curie was almost impossible 30 years ago.

Conclusion: how to fall from normalcy?

There are many reasons why decolonising museums, science centres, and science communication is not as easy as just returning looted artefacts.

For one, dominant Eurocentric knowledge systems are epistemically imposing. The aforementioned case studies demonstrate why they are imposing and how museums and science centres have had to act beyond common knowledge and acceptance to illuminate the contribution of women, women of colour, and LGBTQA+ to science. These case studies are just a very tiny fraction of the massive archive of diverse and inclusive contributions that have been ignored or undervalued by science communication because they do not originate from the 'dominant' or normative standpoint.

Since the dawn of time, humans have been trying to explain the world and natural phenomena linked to their existence. This resulted in a vast production of knowledge that took place at all times and in all places. The story that science communication must tell is not just an interpretation of a phenomenon. If scientific exploration/experimentation consists of trial and error, then we must also concede that it belongs within the historical context of this world. Thus, as science communicators, we cannot be satisfied with a 'sanitised' account of science.

In *Culture and Imperialism*, Said (1993) did not apportion blame to the Western authors whose writings he analysed because, as he deemed it, 'they were children of their times'. No one can blame science museums, science centres, and science communicators for being part of the society of their time either. What they can rightfully be blamed for is the following – that despite being the intermediate link between the academy and other sites for the production of scientific knowledge and the layperson, they did not display or acknowledge the globally inclusive contributions of diverse peoples, regions, and civilisations of the world to the development of the scientific endeavour until this was deemed acceptable by society, and even then, still within the limited boundaries of this limited acceptance.

Falling from normalcy for science museums, science centres, and science communicators means bringing to the fore disciplines, fields of study, and transformative narratives that are not usually brought to the curators' table. Falling from normalcy is to display and interpret objectively contextualised state-of-the-art science with radical and ground-breaking approaches that leap-frog and even precede public opinion and societal times for the greater good and transformation of society at large. Falling from normalcy is to question the understanding of science and what is communicated from the epistemological level. It also entails the integration of diverse global epistemological orientations, including lay epistemologies, into what is considered the 'mainstream' of scientific knowledge.

In their article on 'epistemological orientations', Medin and Bang (2014) define the concept as follows:

> Epistemological orientations also known as lay epistemologies, can be informally defined as the different ways in which people view, conceptualize, and engage with the world. These variations implicitly and explicitly affect knowledge construction and forms of engagement with the world. … In short, epistemological orientations correspond to different ways of seeing the world, each of which may be useful and accurate in itself, but each also providing a different perspective. (Medin and Bang, 2014, p 1)

Integrating lay epistemology implies embracing, acknowledging, and understanding other normalities. Museums and science centres in the West should display and acknowledge science from other civilisations and their corresponding epistemic traditions and worldviews without framing them from the dominant normalcy as a crucial evolution towards decolonisation and transformation in these institutions and in the field of science communication.

Notes

1. http://www.mariatherezaalves.org/works/museum-of-european-normality?c=18
2. https://icom.museum/en/news/icom-announces-the-alternative-museum-definition-that-will-be-subject-to-a-vote/
3. https://icom.museum/en/resources/standards-guidelines/museum-definition/
4. http://www.columbia.edu/itc/anthropology/schildkrout/6353/client_edit/week11/schildkrout.pdf
5. https://globalnews.ca/news/3058929/royal-ontario-museum-apologizes-for-1989-into-the-heart-of-africa-exhibit/
6. https://africasacountry.com/2014/10/brett-bailey-the-barbican-and-black-britons
7. https://hyperallergic.com/436285/casual-colonialism-lara-croft-indiana-jones/
8. http://x3productions.ca/indiana-jones-en [https://www.youtube.com/watch?v=wi6U5zkWpTo&t=4s]
9. https://theconversation.com/racism-is-behind-outlandish-theories-about-africas-ancient-architecture-83898
10. https://www.frieze.com/article/where-do-we-go-here-coffee-care-and-black-panther

References

Baldwin, J. (1998) *Collected Essays: Notes of a Native Son/Nobody Knows My Name/The Fire Next Time/No Name in the Street/The Devil Finds Work/Other Essays*, New York: Library of America.

Baldwin, J. (2010) *The Cross of Redemption: Uncollected Writings*, ed. Randall Kenan, New York: Pantheon Books.

Barthes, R. (1957) *Mythologies*, Paris: Éditions du Seuil.

Bourdieu, P. and Darbel, A. (1966) *L'amour de l'art: les musées et leur public*, Paris: Ed. de Minuit.

Cervulle, M. (2017) 'Exposer le racisme: Exhibit B et le public oppositionnel/ Exhibiting racism: Exhibit B and the oppositional public', *Études de communication*, 48: 37–54. doi: https://doi.org/10.4000/edc.6775

Du Bois, W.E.B. (1910) 'The souls of White folks', in W.E.B. Du Bois (1987), *Writings*, New York: Library of America, pp 923–38.

Fabian, J. (1983) *Time and the Other: How Anthropology Makes Its Object*, New York: Columbia University Press.

Fanon, F. (1952) *Peau noire, masques blancs*, Paris: Éditions du Seuil.

Fanon, F. (1963) *The Wretched of the Earth*, New York: Grove Press.

Foucault, M. (1976) Lectures at the *Cours au Collège de France: Society Must Be Defended (1975–1976)*, New York: St Martin's Press.

Medin, L. and Bang, M. (2014) 'The cultural side of science communication', PNAS Early Edition, Available from: https://psychology.northwestern.edu/documents/faculty-publications/medin-%20communication.pdf

Memmi, A. (1965) *The Colonizer and the Colonized*, London: Earthscan Publication, 2003.

Michael, M. (2016) *Standard White: Dismantling White Normativity*, California Law Review, Inc. https://californialawreview.org/print/3standard-white-dismantling-white-normativity/

Mills, C.W. (1997) *The Racial Contract*, Ithaca, NY: Cornell University Press.

Ngũgĩ, wa Thiongo (2011) *Decolonising the Mind: The Politics of Language in African literature. 1. African Literature – 20th-Century History and Criticism*, Martlesham: James Currey.

Pomian, K. (2020) *Le Musée, une histoire mondiale, T.1: Du trésor au musée*, Paris: Gallimard.

Pomian, K. (2021) *Le Musée, une histoire mondiale, T.2: L'ancrage européen*, Paris: Gallimard.

Price, S. (2009) 'Cultures en dialogues: options pour les musées du XXIe siècle' In *Histoire de l'Anthropologie, Coédition INHA*, Paris: Musée du Quai Branly.

Roediger, D.R. (1999) *The Wages of Whiteness: Race and the Making of the American Working Class*, London: Verso.

Said, E. (1978) *Orientalism*, New York: Pantheon Books.

Said, E. (1993) *Culture and Imperialism*, Vintage digital.

Sarr, F. and Savoy, B. (2018) 'The restitution of African cultural heritage: toward a new relational ethics', report to the French Republic, no 2018–26.

Schildkraut, E. (1991) 'Ambiguous messages and ironic twists: Into the Heart of Africa and the Other Museum', *Museum Anthropology*, 15(2): 16–23.
Spivak, G.C. (1988) *Can the Subaltern Speak?*, Basingstoke: Palgrave Macmillan.
Ünsal, D. (2019) 'Positioning museums politically for social justice', *Museum Management and Curatorship*, 34 (6), Available from: https://www.tandfonline.com/toc/rmmc20/34/6
Vincent, I. (1990) 'U.S. museums cancel Africa exhibit: ROM may lose $100,000 after demise of North American tour', *The Globe and Mail*, 29 November.
Wekker, G. (2016) *White Innocence: Paradoxes of Colonialism and Race*, Durham, NC: Duke University Press.

11

African Challenges and Opportunities for Decolonised Research-Led Innovation and Communication for Societal Transformation

Akanimo Odon

Introduction

The role of decolonised science, technology, and innovation

In an era where there is a crucial need to identify long-term solutions to sustainable development and global change challenges across the world, there is a strategic lead that science, technology, and innovation (STI)-focused organisations in Africa (including governments, universities, and research and development [R&D] departments of industries) can take to develop homegrown policies and initiatives to allow Africans themselves to provide solutions to their own particular needs and challenges.

The African continent is bedevilled by multiple challenges that cut across different economic sectors. In the energy sector, 600 million people in sub-Saharan Africa are without electricity access – this is about 60.7 per cent of the population, and 50 per cent of businesses view a lack of reliable electricity access as a major constraint to doing business (IEA, 2019). There are also problems of access to clean and safe water: 30 per cent of Nigerians (57 million people), 10.7 per cent of Ghanaians (3 million people), and 36 per cent of Kenyans (17 million people) have no access to clean water. In the same vein, 68 per cent of Nigerians (130 million people), 78 per cent of Ghanaians (22 million people), and 67 per cent of Kenyans (32 million people) do not have access to improved sanitation (UNICEF, 2021; World

Bank, 2021). Challenges also abound in Africa and across other regions of the Global South when it comes to issues of sustainable development, such as food security, gender equality, and education management. However, these challenges also provide opportunities for critical interventions through decolonised research and innovation undertaken by science and technology organisations (AUC, 2014; AUC, 2015b; African Union, 2019; African Union, 2020).

However, though many African universities that undertake research and teaching and research-only agencies and organisations spend a lot of time and resources doing research, there is an insufficient drive to provide solutions to these challenges. This seems to be the case in formal research settings where there is an overwhelming push for publication of international research papers as a basis for promotion and career enhancement. In the non-traditional research settings devoid of popular institutional structures, there is constant research to offer solutions to societal problems, but this has only had a limited impact. These problems have grown over generations due to a combination of multiple factors, including accumulative failings by national governments, poor leadership, limited investment for development, and inconsistent policy frameworks. Communication of this 'bottom-up' research-led innovation agenda has been undermined by communication barriers and a lack of alignment with frameworks set by the Global North. This points to the need for a decolonised framework in order to upscale and transform this Afrocentric approach (Asante, 2011).

A typical example is Africa's traditional medicine sector. Traditional medicine has been used for the health and well-being of Africans for many generations but has only started getting some recognition and 'mainstreaming' in the last 20 years. More than 40 African countries now have traditional medicine policies, many of which have been integrated into national health policy frameworks (WHO, 2022). Most African countries also have policies to regulate traditional medicine practitioners. Universities in about 25 countries in Africa now offer traditional medicine courses to their pharmacy and medical students, and in up to 17 countries traditional medicine and contemporary health provision have been integrated, even to the point of providing referral pathways between these two paradigms (WHO, 2022). Three African countries (Mali, Ghana, and South Africa) have even gone as far as establishing partial health insurance coverage for traditional medicine products and services. There is still the issue of the exportation of these traditional medicine products, but then they have to comply with international standards set by the Global North, and notwithstanding the imperatives of international intellectual property rights, which limit the capacity for international trade. With over 30 research institutes now dedicated to traditional medicine in Africa, there is an opportunity for increased productivity, efficiency, and transformative Afrocentric innovation in the sector, based on Africa's Indigenous knowledge systems, traditions, and cultural sensitivities and science communication structures (WHO-Regional Committee for Africa, 2013; AUC, 2015a; AUC, 2015b; WHO, 2022).

Most of the progress that has been made in this sector has been driven by national governments and institutions taking ownership and responsibility for the communication of their innovations in radical and progressive alignments with frameworks set up by the Global South. Although these local systems and progressive sociocultural frameworks require optimisation to further enhance quality, delivery systems, and dosages backed up by research, it is encouraging that progress is being made to deliver on the promise of a decolonised health sector for the African continent.

The COVID-19 pandemic ushered in a global race for technological leaps in the speed of development and delivery of the COVID-19 vaccines in which Africa and most of the Global South regions lagged behind. The role of science, technology, and innovation represented by the capacity to provide solutions to societal problems remains vital in the decolonisation of research, and unless Africa enhances its position and profile in the areas of its local research, it will continue to be marginalised (Hountondji, 1990; Boidin et al, 2012; AUC, 2020). There was and remains a global debate on the unequal access of Global North and Global South regions to the COVID-19 vaccines and the inequality and marginalisation of the Global South that frustrates the role of science and technological innovation in global transformations.

Sustainable and equitable partnerships for transformation

There are undoubtedly huge gaps between academic research, industry, and government (the triple helix) in many African countries that require strategic interventions. There is an urgent need to design innovative and decolonised policy tools, knowledge transfer programmes, and pathway-to-impact models to engender a new era of impact-oriented research and public engagement in African countries, driven by STI-backed solutions to problems (Mugwagwa et al, 2018). This strategic push is needed to support the process of change from a 'colonised' legacy to a 'decolonised and ethical' agenda in African research. However, for this transformation to happen, there is a need to bridge the gaps between the triple helix of strategic stakeholders. One of the challenges that stifles the implementation of decolonised STI policies, research, and programmes and their effective communication with African publics is the silos-working mentality of the tripartite segments of society as defined earlier.

The communication of science using language constructs and systems that cut across the triple helix of academia, industry, and government is a fundamental component of a transformational and decolonised landscape. However, over the years, the gaps between academia and other segments have widened due to the elitist status that researchers have assumed devoid of connection with local industry or communities. Viewed through a quadruple helix model, in which local societies and communities are also a crucial segment, it is clear that they remain the least engaged and communicated to, even though they are supposed to be the most important beneficiaries of scientific solutions. Currently, the communication of science has been restricted to fellow academic researchers who understand scientific jargon, thus further widening the gap that already exists between African researchers/scientists and the African publics. However, since it is crucial that the triple helix segments work seamlessly together for national development, the question arises of who should lead that conversation for a connection between them. Who should ensure that they work together in partnerships that will deliver the sustainable development fruits of publicly engaged scientific research within a decolonised framework?

Another critical dimension to the challenges of delivering decolonised science communication in the African context, as in much of the Global South, is the 'colonisation of research' dressed up in the guise of collaborative international partnerships. There are a range of issues that make these sorts of partnerships untenable and unsustainable and in some respects blatantly unethical, particularly when the dominant partner in the Global North shapes the research agenda, framework, and even its core focus to suit its own purposes without due consideration of local African circumstances, contexts, and priorities (Kok et al, 2017). In some cases, these international partnerships are based on providing an innovative solution to a local challenge in health-related issues such as infectious diseases, yet the resultant action is often a research publication or an innovation asset that assumes that the dominant northern partner/institution alone knows the problems and is solely capable of providing the solutions. This is done without due respect and recognition of the critically important role and contribution of the local African partner/institution.[1] This is science and research colonialism. In some other cases, local African partners are simply used as data collectors for major internationally funded research projects without due opportunities for co-authorship of research outputs, marginalising the local partners and colonising their research (Shose et al, 2020).

For sustainable international collaboration partnerships to be equitable, it is imperative to view these partnerships through a decolonisation lens, which recognises that there are differences between partners in the Global North and the Global South. These differences should complement each other, making research more impactful, transformational, and publicly engaged. The differences in roles, culture, benefits, power dynamics, capabilities, and motivations should be discussed and deliberated upfront at the conceptualisation stages of these international partnerships and compromises reached before programmes are initiated. Partners/institutions in the Global South should be able to challenge the status quo and request equity in the delivery of research projects and their decolonised communication in these international collaborative partnerships.

International funding programmes should take note of these unequal power dynamics and mandate equitable partnerships for international research project funding applications. The more that individuals and institutions from the Global South lead and take ownership of these research projects, the greater the opportunities to enhance science communication devoid of colonised premises and ideologies (Ozor, 2015).

The funding divide and bridging the gaps

There is an underlying pressure on researchers in Africa to communicate science in specific ways that align with expectations and frameworks set by research funders and proponents of the internationalisation of the science communication field in the Global North. This opens up a conversation that justifies the need for the decolonisation of the research and science communication agenda, especially if it must attend to local challenges and create developmental opportunities. Pressure to communicate science to externally imposed standards set by the Global North – which is often linked to research career promotion and funding – restricts the drive for science communication for local benefit in Africa.

However, eco-innovation and decolonisation cannot be achieved without qualitative science and technology-focused research and resource mobilisation, which must be driven by government policies to bridge the gaps between research and society. Even with all the critical policies and strategies in place, implementation would still struggle in the absence of funding.

The African Union's ten-year (2014–24) Science, Technology, and Innovation Strategy for Africa (AUC, 2014) emphasises that African states have the potential to make a 1 per cent contribution of their Gross Domestic Product (GDP) towards financing STI programmes that are led by African STI-focused organisations. African leaders have always committed to increasing funding for national, regional, and continental programmes for science and technology, but so far only a handful of countries have implemented that pledge. Although there is progress in some countries, with R&D budgets as a ratio of GDP close to the 1 per cent target in a few, most countries on the continent remain below the 0.5 per cent mark, with some countries still at the 0.1–0.2 per cent mark (AUDA-NEPAD, 2019). In 2013, Africa's gross expenditure on R&D was about 0.45 per cent of GDP compared with 2.71 per cent in North America, 2.10 per cent in South East Asia, 1.75 per cent in Europe, 1.62 per cent in Asia, and 1.03 per cent in Latin America and the Caribbean (AfDB, 2019a). It is thus clear that, irrespective of the availability of external funding, African governments need to invest more in the production and dissemination of scientific knowledge that is fit for purpose and advances the decolonisation of the scientific enterprise and its communication with the public on the continent. Despite improvements in some countries, much greater levels of public investment in science and its communication – and, by extension, ownership of and decolonisation of local science agendas – will be required if the ideal of intellectual decolonisation is to become a reality in Africa.

Africa is home to just 2.5 per cent of the world's researchers (1.1 per cent for sub-Saharan Africa and 1.4 per cent in North Africa), compared with 42.8 per cent in Asia, 31 per cent in Europe, 18.5 per cent in North America, and 3.6 per cent in Latin America and the Caribbean (AfDB, 2019a, 2019b, 2019c). In most African countries, STI-focused research institutions depend wholly on their national governments, which are now struggling to fund them.

In many African countries, more than half of the investment and funding into higher education research is from international donor funds, which in itself places African institutions at a disadvantage when it comes to negotiating local content and context in research and decolonised science communication agendas (Bendana, 2019). Several financing mechanisms and models have been crafted to spur the growth of technological innovations that are needed to contribute and support higher education, science, technology, innovation, and R&D in creating employability, skills, and productivity in the labour market (Mugwagwa et al, 2019; UNEP 2016). All over the world, there have been major transformations in higher education funding, but there have been clear differences between the Global North and Global South. While funding for education and research in the Global North is spread relatively evenly between public and private funding sources, minimal funding is provided by the private sector in Africa, meaning that African researchers are overwhelmingly dependent on government funding, which is limited in the first place. This further places African institutions in a position of subservience to their international funding partners and impedes the decolonisation agenda in scientific research and its communication for transformational impact (Panigrahi, 2018). It is therefore vital that African institutions intensify their strategic partnerships with the private sector on the African continent to enhance their localised funding regimes and reduce their dependence on the public sector and on international donor funding.

Opportunities for decolonising science communication

Does science have a unique language? This is a question that is being asked whether intentionally or inadvertently. The benefit of decolonised science communication is that it offers different perspectives to solving practical global problems within the framework of multi- and trans-disciplinarity. This means that local underrepresented and Indigenous people must be an important stakeholder group in the articulation of the science agenda, project development, and delivery, because that will also influence science communication to those groups (Aikenhead, 2001). Decolonised science communication builds trust and acceptance because of the premise of ownership it conveys and establishes (Rasekoala, 2022).

The problem of the single story described by the African writer Chimamanda Ngozi Adichie (2009) should be avoided in science communication. It allows for colonised perceptions and theories that create stereotypes and accordingly a premise that maybe true but most likely would be incomplete. There are other points of view, and this is the basis of decolonised science communication that inclusively allows for other languages and cultures (Nolwazi, 2015).

There is a need to avoid biased labelling that enforces colonised constructs in science communication. In a chapter on Chinese medical doctors in Kenya, Hsu (2012) critiques medical anthropological studies that focus on illness narratives while ignoring those of practitioners. She points out that complementary and alternative medicines exist in the Global North, but when such practices exist in the Global South, they are labelled 'traditional medicine', damaging their reputation and global acceptability. Why aren't the same practices given the same label in both the Global North and South regions?

The global effort to meet the transformational capacity of aligning and complying with the United Nations Sustainable Development Goals (SDGs) offers an opportunity to consider decolonising science communication as a crucial tool to support the achievement of these goals. The 17 SDGs focus on the development of people, prosperity, peace, and the planet through strategic and equitable partnerships (UNGA, 2015). It is therefore impossible to fulfil the objectives of the UN SDGs if the communication of their underlying ethos is skewed within a colonised system.

Evidence of good communication is an assurance of understanding and clarity from that communication both by the speaker and the listener. In an interview, Sibusiso Biyela, a science communicator and journalist in South Africa, articulated his struggle that while he can speak to his close friends and family about a whole host of topics in his mother tongue, isiZulu, he is forced to switch to English when talking about science. Like most Africans, he further emphasises that he is quite good at explaining science concepts to non-scientists but quickly reaches his limits when doing so in his mother tongue. He concludes with a question – can you really understand science concepts if you can't explain them in your language (Pichon, 2021)?

Conclusion

The advance of inclusive science communication for societal transformation in the Global South regions of the world would be considerably strengthened by consciously leveraging and integrating the decolonisation agenda into global research. Doing so would address critical questions of equity, social justice, peace, and democracy. In the African context, the case has been made for the development of ethical protocols centred on the communal and relational ties that are highly valued by people in those societies (see Seehawer, 2018; Keikelame and Swartz, 2019). In an environment where challenges abound, in which innovation-driven research can proffer practical solutions to societal problems, there is a profound need for an evolution in decolonised research culture to adapt to changing times globally and sustainable development imperatives on the African continent.

Researchers and academics in Africa should undertake and communicate research that has been developed on the basis of local conditions, cultural dimensions, and their geographical peculiarities. In the same vein, science and innovation should be carried out under this pretext for sustainability. Empowering concepts such as co-design, co-application for funding, co-delivery of projects, and the co-hosting of science projects need to be the only way by which collaborative research partnerships between the Global North and African researchers/institutions are administered in order to advance decolonisation and engender impactful societal transformation. In addition, in much the same way as there is a growing demand and global mandatory requirement to include gender balance and equality strategies in research funding applications, a similar requirement should be mandated for the inclusion of a decolonised science communication strategy for funded project effectiveness and sustainability. This is particularly crucial in multi-partner science research projects involving partners from the Global North and South. An even greater performance management and quality assurance measurement of the project should be the requirement to show evidentially how truly 'decolonised' the project has been.

The role of science, technology, and innovation represented by the capacity to provide solutions to societal problems remains vital in the decolonisation of research and its culturally centred communication in Africa. It is time for African researchers, institutions, and science communicators to strategically enhance their positions and profiles at institutional, national, and international levels.

The process of decolonising science communication must be intentional and include the diversification of the sources being engaged or quoted, the use of more inclusive languages, and diverse cultures. There is a need to ensure local content within scientific research projects, which should begin with the assessment of cultural paradigms and historical contexts at the project conceptualisation stages. The intention should be to build a decolonisation strategy within the operationalisation of the project to right the wrongs wherever they exist. For the future of the African continent and to correct the mistakes of the colonised past and their enduring legacy, the mentoring and training of young African researchers in new decolonised models of science communication is absolutely crucial and provides an opportunity for progress (Karikari et al, 2016). The #MeToo and the Black Lives Matter movements have attempted to challenge the status quo. Is it not time for a science decolonisation movement that does the same? In the virtually connected world in which we now live, it is important for science communication to be easily accessible in decolonised constructs and forms that diverse publics fully understand.

Note

[1] https://mg.co.za/health/2022-06-13-the-lancet-journal-rejects-papers-that-dont-acknowledge-african-researchers/

References

Adichie, C. (2009) 'The danger of a single story', TED Talk, Available from: https://www.ted.com/talks/chimamanda_ngozi_adichie_the_danger_of_a_single_story?%20language=en

African Development Bank Group (AfDB) (2014) 'African Development Bank's human capital strategy 2014–2018', Available from: https://www.afdb.org/fileadmin/uploads/afdb/Documents/Policy-Documents/AfDB_Human_Capital_Strategy_for_Africa_2014-2018.pdf

African Development Bank Group (AfDB) (2019a) 'Unlocking the potential of the fourth industrial revolution in Africa study report', Available from: https://4irpotential.africa/wp-content/uploads/2019/10/4IR_NIGERIA.pdf

African Development Bank Group (AfDB) (2019b) 'Creating decent jobs', Available from: https://www.afdb.org/en/documents/creating-decent-jobs-strategies-policies-and-instruments

African Development Bank Group (AfDB) (2019c) 'Creating decent jobs instrument and policy tool', Economic Outlook Report, Available from: https://am.afdb.org/2019/sites/default/files/AfDB18-16_Jobs_English.pdf

African Union (2019) 'Africa's STI implementation report', Available from: https://au.int/sites/default/files/newsevents/workingdocuments/37841-wd-stisa-2024_report_en.pdf

African Union (2020) 'African economic outlook: developing Africa's workforce for the future', Available from: https://au.int/sites/default/files/documents/38116-doc-african_economic_outlook_2020_.pdf

African Union Commission (AUC) (2014) 'Science, technology and innovation strategy for Africa 2024', Available from: https://au.int/sites/default/files/newsevents/workingdocuments/33178-wd-stisa-english_-_final.pdf

African Union Commission (AUC) (2015a) 'Agenda 2063: "the Africa we want"', Addis Ababa: AUC.

African Union Commission (AUC) (2015b) 'Agenda 2063, "first ten-year implementation plan 2014–2023"', Available from: https://au.int/sites/default/files/documents/33126-doc-11_an_overview_of_agenda.pdf

African Union Commission (AUC (2020) 'The digital transformation strategy for Africa (2020–2030)', Available from: https://au.int/en/documents/20200518/digital-transformation-strategy-africa-2020-2030

African Union Development Agency (AUDA-NEPAD) (2019) Annual Report. Available from: https://www.nepad.org/publication/auda-nepad-2019-annual-report

Aikenhead, G.S. (2001) 'Integrating Western and aboriginal sciences: cross-cultural science teaching', *Research in Science Education*, 31: 337–55.

Asante, M.K. (2011) 'De-Westernizing communication: strategies for neutralizing cultural myths', in G. Wang (ed) *De-Westernizing Communication Research: Altering Questions and Changing Frameworks*, New York: Routledge, pp 21–7.

Bendana, C. (2019) 'African research projects are failing because funding agencies can't match donor money', *Science*, Available from: https://www.sciencemag.org/news/2019/04/african-research-projects-are-failing-because-funding-agencies-can-t-match-donor-money

Boidin, C., James C., and Ramon, G. (2012) 'Introduction: from university to pluriversity: a decolonial approach to the present crisis of Western universities', *Human Architecture: Journal of the Sociology of Self-Knowledge*, 10(1): 1–6.

Hountondji, P. (1990) 'Scientific dependence in Africa today', *Research in African Literatures*, 21(3): 5–15.

Hsu, E. (2012) 'Mobility and connectedness: Chinese medical doctors in Kenya', in H. Dilger, A. Kane, and S.A. Langwick (eds) *Medicine, Mobility, and Power in Global Africa: Transnational Health and Healing*, Bloomington: Indiana University Press, pp 295–315.

International Energy Agency (IEA) (2019) 'Africa energy outlook: special report', Available from: https://www.iea.org/reports/africa-energy-outlook-2019

Karikari, T.K., Yawson, N.A., and Quansah, E. (2016) 'Developing science communication in Africa: undergraduate and graduate students should be trained and actively involved in outreach activity development and implementation', *Journal of Undergraduate Neuroscience Education*, 14(2): E5–E8.

Keikelame, M.J. and Swartz, L. (2019) 'Decolonising research methodologies: lessons from a qualitative research project, Cape Town, South Africa', *Global Health Action*, 12:1561175. doi: 10.1080/16549716.2018.1561175

Kok, M.O., Gyapong, J.O., Wolffers, I., Ofori-Adjei, D., and Ruitenberg, E.J. (2017) 'Towards fair and effective North–South collaboration: realising a programme for demand-driven and locally led research', *Health Research Policy and Systems*, 15(96). doi: https://doi.org/10.1186/s12961-017-0251-3

Mugwagwa, J., Banda, G., Mjimba, V., Mavhunga, C., Muza, O., Bolo, M. et al (2018) 'Unpacking policy gridlocks in Africa's development: an evolving agenda', *International Journal of Technology Management and Sustainable Development*, 17(2): 115–34.

Mugwagwa, J., Banda, G., Ozor, N., Bolo, M., and Oriama, R. (2019) 'Optimising governance capabilities for science, research and innovation in Africa', IDRC Grant: 108460-001-Networking Africa's science granting councils.

Nolwazi, M. (2015) 'Medical anthropology in Africa: the trouble with a single story', *Medical Anthropology*, 35(2): 193–202. doi: 10.1080/01459740.2015.1100612

Ozor, N. (2015) 'Increasing opportunities for financing of research: lessons for the demand side; international symposium on "new science, technology and innovation funding mechanisms in Africa"', 9–11 December in Dakar, Senegal.

Panigrahi, J. (2018) 'Innovative financing of higher education', *Higher Education for the Future*, 5(1): 61–74. doi: 10.1177/2347631117738644

Pichon, A. (2021) 'The language of science', *Nature Chemistry*, 13: 1025–6. doi: https://doi.org/10.1038/s41557-021-00822-y

Rasekoala, E. (2022) 'Responsible science communication in Africa: rethinking drivers of policy, Afrocentricity and public engagement', *JCOM: Journal of Science Communication*, 21(4): C01. doi: https://doi.org/10.22323/2.21040301

Seehawer, M.K. (2018) 'Decolonising research in a sub-Saharan African context: exploring Ubuntu as a foundation for research methodology, ethics and agenda', *International Journal of Social Research Methodology*, 21(4): 453–66. doi: 10.1080/13645579.2018.1432404

Shose, K., Zoe M., and Elelwani, R. (2020) 'Decolonising African studies', *Critical African Studies*, 12(3): 271–82. doi: 10.1080/21681392.2020.1813413

United Nations Environmental Programme (UNEP) (2016) 'Is Africa's natural capital the gateway to finance its development?', Available from: https://www.unep.org/news-and-stories/story/africas-natural-capital-gateway-finance-its-development

UNGA (2015) 'Transforming our world: the 2030 Agenda for Sustainable Development', UN General Assembly, New York, Available from: https://sdgs.un.org/2030agenda#:~:text=We%20resolve%2C%20between%20now%20and,protection%20of%20the%20planet%20and

UNICEF (2021) 'Nearly one third of Nigerian children do not have enough water to meet their daily needs', press release, Available from: https://www.unicef.org/nigeria/press-releases/nearly-one-third-nigerian-children-do-not-have-enough-water-meet-their-daily-needs

World Health Organization (WHO), Regional Committee for Africa (2013) 'Enhancing the role of traditional medicine in health systems: a strategy for the African region', document AFR/RC63/6, WHO, Regional Office for Africa.

World Health Organization (WHO) (2022) 'Africa Traditional Medicine Day 2020', Available from: https://www.afro.who.int/regional-director/speeches-messages/african-traditional-medicine-day-2020

World Bank (2021) 'Feature story: Nigeria; ensuring water, sanitation and hygiene for all', Available from: https://www.worldbank.org/en/news/feature/2021/05/26/nigeria-ensuring-water-sanitation-and-hygiene-for-all

12

Decolonising Science Communication in the Caribbean: Challenges and Transformations in Community-Based Engagement with Research on the ABCSSS Islands

Tibisay Sankatsing Nava, Roxanne-Liana Francisca, Krista T. Oplaat, and Tadzio Bervoets

Introduction

Effective public engagement and science communication are some of the cornerstones of translating and applying science into real-world applications. Whether it is in terms of communicating the efficacy of vaccines or the role of protected areas in biodiversity conservation efforts, it is a critical yet often neglected component of those involved in the field of STEM (science, technology, engineering, and mathematics). 'Science and engineering lack a culture of explanation' (Meredith, 2010, p 6), and this is further compounded when additional factors are considered. Scientists often ascribe to the myth of universal applicability of their research and fail to consider historical and sociocultural complexities (Meredith, 2010; Mbembe, 2016). They appear 'arrogant and aloof when talking about their subject, especially when discussing with disenfranchised communities' (Olson, 2009), while those communities have relevant knowledge and perspectives for science and could often benefit from integrating peer-reviewed scientific results to guide policy decisions. Orthia (2020) argues that 'science communicators must take steps to radically reform their understandings of [science communication] so that people from diverse cultures, nations and traditions can genuinely own it as theirs'.

Science communication and public engagement in the ABCSSS islands

The six Caribbean islands of Aruba, Bonaire, Curaçao, Saba, Sint Eustatius, and Sint Maarten are part of the Kingdom of the Netherlands. The islands are collectively referred to as the ABCSSS islands. In the ABCSSS islands, contemporary science communication initiatives of STEM research mainly engage White, highly educated, Dutch- or English-speaking audiences. This echoes similar findings in Europe (Dawson, 2014) and can be traced directly to who funds, designs, leads, executes, communicates, and benefits from scientific research and its results. In the ABCSSS islands, many science communication activities are carried out according to the 'deficit' model, which assumes a lack of knowledge in the target audience that can be remedied by unilateral, top-down communication of research goals, processes, methodologies, and results (Horst et al, 2017; Burns, 2018). While some important exceptions are highlighted in this chapter, STEM initiatives that prioritise public engagement that is participatory, reciprocal, and community-based throughout the research process are less common (Palmer and Schibeci, 2014; Horst et al, 2017; Sankatsing Nava and Hofman, 2018). In contrast, many locally led social science and humanities projects in the Caribbean are deeply grounded in and informed by community-based research and engagement (Allen, 2018; CaribResearch Research Agenda, 2022).

Research, funding, and science communication infrastructures of the ABCSSS islands

Since 10 October 2010, the Kingdom of the Netherlands consists of four autonomous countries: Aruba, Curaçao, Sint Maarten, and the Netherlands. Besides the European territory, the Netherlands includes three islands in the Caribbean region: the 'special' municipalities Bonaire, Saba, and St. Eustatius. There are a variety of organisations and individuals involved in and various approaches to science and science communication in the Caribbean part of the Kingdom of the Netherlands. Aruba, Curaçao, and Sint Maarten have their own universities, and each island has its ecosystem of (independent) researchers and research and higher education institutes and organisations. The knowledge centres are small and have (comparatively) small budgets, prioritise education over research, are only partly locally staffed, and generally do not have specialised science communication departments. At the same time, these organisations have an important role in engaging island communities with science. ABCSSS researchers are dependent on foreign universities, foreign research funding bodies, and ad-hoc government funding. To illustrate, it is only since 2019 that researchers and universities in the Caribbean islands of the Kingdom of the Netherlands have qualified for funding from the Dutch Research Council to lead their own research projects (NWO, 2019). And even when islanders qualify, the conditions attached to international funds are often so specific that it is virtually impossible for locally based non-governmental organisations and researchers to succeed independently.

One often overlooked factor in science and its communication is the role that Black, Indigenous, and People of Colour (BIPOC) communities play in integrating scientific results to effectively foster positive community change. Science communication often excludes the geopolitical, socio-economic, and cultural realities of the communities in the ABCSSS islands. This echoes Saran Stewart's observation that 'common to the Caribbean is an understanding of how colonial legacies of research have ridiculed oral traditions, language, and ways of knowing, often rendering them valueless and inconsequential' (Stewart, 2019, p ix). These colonial legacies and the particular entanglements of colonialism, scientific research, nature conservation, and (mental) health care in the Caribbean are reflected in significant challenges in public engagement with science, and are key themes tackled in this chapter.

Decolonising science communication means decolonising science

This chapter reflects on opportunities to decolonise science communication in the ABCSSS islands. With this in mind, the authors (whose biographies in this book include a brief positionality statement) began by considering the word 'decolonising' tentatively with regard to science communication, because decolonisation is not an umbrella term for all social justice, anti-racism, or diversity and inclusion efforts. In fact, decolonisation 'is not a metaphor' for actions that do not 'bring about the repatriation of Indigenous land and life', and cannot be used for all the positive changes we want to make in our societies (Tuck and Yang, 2012). Decolonisation *unsettles* and is uncomfortable. Thus, when speaking about decolonising science communication, the chapter refers to what Bagele Chilisa defines as self-reflectively 'centring the concerns and worldviews of the colonised Other so that they understand themselves through their own assumptions and perspectives' (Chilisa, 2019). The authors use Chilisa's understanding of what 'decolonising' can mean to explore some pitfalls of science communication in the Caribbean and share examples that amplify voices and centre the needs of those who are excluded or unheard. Throughout this chapter, the focus is on science communication; however, the analysis also touches upon science and research practice more generally. This is inevitable, as it is impossible to consider 'decolonising' science communication without critically reflecting upon the structures and practices that academic research in the Caribbean is built on and continues to perpetuate.

Using this theoretical lens, the authors illustrate the challenges described in this introduction through an in-depth exploration and analysis of two case studies in the fields of nature conservation and mental health care. This chapter focuses on these two disciplines and as such does not reflect the broad range of social science and humanities research, including community-embedded research projects led by universities and independent researchers in the ABCSSS islands in disciplines of heritage, culture, gender, law, healthcare, and beyond.

The cases highlight the lack of community-engaged research in nature conservation and mental health care in the ABCSSS islands. The analysis of the challenges in nature conservation and mental health care communication lays the groundwork to present transformative practices to build a more embedded engagement with science in the ABCSSS islands. Finally, the chapter concludes with a reflection on the (im)possibilities of decolonising science communication and offers an alternative vision of community-based engagement with science in Aruba, Bonaire, Curaçao, Saba, Sint Eustatius, and Sint Maarten.

Challenges in public engagement with nature conservation in the Caribbean

'Helicopter science' in conservation research

The research and conservation agendas in the ABCSSS islands are primarily dictated by the interests of the European Netherlands and other foreign institutions by way of their access to funding opportunities and technical skills. For example, inhabitants of Bonaire, Saba, and Sint Eustatius were long excluded from Dutch subsidies to accelerate the transition to renewable energy (Milieu Centraal, 2022; Rijksoverheid, 2022). In 2022, the Dutch government reserved the first significant budget for nature and the environment for these islands since they became part of the Netherlands. This shows that there is 'a dependency on post-colonial powers to guide conservation actions of former colonies' in Caribbean nature conservation (Hall and Tucker, 2004). As a result, the emphasis is often on current trends in academia. This makes it difficult to get and maintain long-term support and involvement from the local population. Bonaire is a clear example. Each year, academic research is published about the state of the reef, foraging dynamics of various species, and the island's geology. However, important needs such as fisheries research, stock assessments, and climate change adaptation and mitigation are rarely addressed. It is therefore pertinent to question the extent to which the interests of foreign researchers align with local interests and the topics that most affect daily life for the inhabitants. On top of that, the islands often deal with so-called 'helicopter science', where, once a research plan has been established (often by foreign researchers), researchers fly in, do their research, and leave without engaging with local communities, universities, or government in any meaningful ways. Mac Donald (2022) also highlights this when she describes the wariness of fisherfolk and other stakeholders to collaborate with researchers as they require much time and information and provide little to no follow-up on their findings.

A glimpse into the scientific literature on nature conservation in the ABCSSS islands dating back from 1901 shows the imbalance in the recognition given to on-island knowledge and expertise that is often crucial to the execution of the research. This is a widely recognised phenomenon, for which Indigenous scholar and librarian Lorisia Macleod designed citation templates to 'find a better way to acknowledge [Indigenous] voices and knowledges within academia' (Macleod, 2021). For example, within nature conservation research, the names of 'on-island' contributors are rarely included in the author list, even though it is often their knowledge of local conditions, history, and phenomena being studied that form the basis of most research. This is without considering the expertise that goes into logistics, site selection, navigating the social and political climate, and other intangibles that local researchers and community members contribute, and without which the research would not succeed. While foreign researchers often rely heavily on input in the data collection phase of their work, they rarely request this when deciding on the topics, analysis, or communication of their research. But when they do, the on-island contributors are usually still excluded from the research funding for projects that contain their ideas.

Nature parks and the fortress model in the ABCSSS islands

In 'Dutch' Caribbean nature conservation, science has been used to establish protected areas without the input of local communities, which often results in reduced efficacy of the protected area to enhance biodiversity (Zaitchik, 2018). To illustrate, many of the nature parks in the islands were established without taking local stakeholders into account. This is the case for Washington Slagbaai National Park (1969), Christoffel Park (1978), Bonaire National Marine Park (1979), and Saba National Marine Park (1987) (Dutch Caribbean Nature Alliance, 2021). The resulting model relies on 'fortress conservation' based on the belief that biodiversity protection is best achieved by creating protected areas where ecosystems function in isolation from human disturbance. This model assumes that local people use natural resources in irrational and destructive ways (Rai et al, 2021). Such protected areas exclude local people dependent on the natural resource base through a 'fines and fences' approach enforced by park rangers and consider tourism and scientific research as the only appropriate uses for protected areas (De Santo et al, 2011). These disenfranchising approaches result in conservation conflicts, as local people are labelled as criminals, poachers, and squatters on lands they have historically occupied (Robbins, 2007) and have a negative impact on public engagement with nature conservation on the islands. The establishment of nature parks on the ABCSSS Islands emphasises the continuities between environmentalism and the colonisation of the Caribbean, described eloquently by Martinican thinker Malcom Ferdinand (2021). These analyses of science communication efforts in conservation on the ABCSSS islands are informed by global conversations on decolonising conservation through the work of Indigenous conservationists and other scholars (Connell, 2017; Blair, 2019; Canon, 2019; Zanotti et al, 2020; Ferdinand, 2021; Mabele et al, 2021). Particularly with regard to the example of protected nature areas, there is an element of what Chilisa describes as scientific colonialism:

> [R]esearchers travelled to distant colonised lands where they turned resident people into objects of research. This carried with it the belief that the researchers had unlimited rights of access to any data source and information belonging to the population, and the right to export data from the colonies for purposes of processing into books and articles. (Chilisa, 2019, p 7)

Strategies and language in communicating nature conservation

In nature conservation research in the ABCSSS islands, many research proposals include the words *capacity building* and *increasing awareness*, yet these actions are often an afterthought and are rarely tailored to the local situation. Scientific results are often published in language and media largely inaccessible to the general population. An example from the ABC islands is the lack of communication products produced or presented in the islands' languages (Papiamento/u). It is a simple, yet often overlooked, strategy to understand how local populations communicate before designing any communication or engagement campaign. In many cases, it is not necessary to reinvent the wheel: there are plenty of individuals and organisations 'on island' with experience working with communities that are hard to reach for others. These organisations can support researchers in building their own capacity to plan societally embedded research and engage with island communities effectively. Through an annual student exchange programme, the University of Aruba encourages the development of community-engaged student research. Such locally led training is crucial in building skills and experience in the future research population (Mijts et al, 2022).

Often, local researchers also do not prioritise engaging the public, nor link their research to the existing universities on the islands. Local researchers lack the funding and infrastructure required to support effective engagement. Most researchers are therefore willing to do a public presentation, and some will engage policy makers, but few make efforts beyond this. It is therefore also important for researchers from the islands to reflect on their public engagement activities. Being local does not automatically make researchers good community partners or communicators, and the divide between academics and non-academics remains within Caribbean society as well. In contrast, Brenchie's Lab is a community maker space in Aruba that organises long-term citizen science initiatives in which community members are actively involved in environmental research and monitoring. These initiatives support communities to collect environmental data themselves, to initiate research projects, and identify local issues to be included in the global conversation (Sevold, 2020). Brenchie's Lab projects include collecting beach sand samples to measure microplastics, mapping coastal changes using Google Earth, and measuring ocean acidity together with islanders. In the ABCSSS islands, however, the governments often rely on the insights of foreign academics rather than those of local experts. This remains a challenge, especially for local organisations that involve non-academic communities in research or collect data in non-traditional ways.

Despite these challenges, there is an increased and concerted effort to raise awareness, involvement, and ownership in conservation among island residents. On the islands, there are various programmes that focus on community engagement and involvement (DCNA, 2021). One example is the use of emblematic species that have a cultural and historical significance to local populations to frame conservation messaging. In 2011, the Sint Maarten Nature Foundation launched its Pelican Conservation programme. Despite there being no significant pressure on the species, this programme engaged island populations in citizen science to build ownership for a national symbol and its associated habitat. Participants were not only introduced to species-specific conservation and the conservation of associated habitat but were also sensitised to the environmental pressures on said habitat, resulting in the establishment of an important Bird and Biodiversity Area for one of the locations monitored. The communication of this project was done on a community level, with community-focused dialogues, presentations, and dissemination using traditional and social media (Sint Maarten Nature Foundation, 2011). The fisheries cooperative PISKABON on Bonaire is another example. The cooperative was established in 2017 to actively involve local fishermen in the management procedures of the marine environment of Bonaire. By investing in the fisher community through the cooperative, local fisherfolk regain a sense of ownership and responsibility of the marine environment (Mac Donald, 2022).

Mental health care communication in Bonaire, Saba, and Sint Eustatius

Mental health care on the ABCSSS islands is confronted with similar challenges as described for nature conservation. For example, organised mental health care in Bonaire, Saba, and Sint Eustatius has been financed by the Dutch Ministry of Health since 2010 and was reshaped and expanded from the former foundations to the current Mental Health Caribbean Foundation. This foundation was initiated by a Dutch institute for mental health in close cooperation with the Dutch academic medical centres of the Vrije Universiteit Amsterdam/UMC, which also provide health care specialists. Similar constructs with Dutch (academic) institutions are also found in Aruba, Curaçao, and Sint Maarten, in both somatic and mental health care.

Concurrently, the rich and varied cultural intricacies of the ABCSSS islands also carry with them Afro-Caribbean beliefs, rituals, and spiritual elements, including Indigenous and Euro-Christian religious practices, that offer traditional healing methods for mental illnesses, such as Brua or Obeah (Blom et al, 2015). Despite being poorly researched, these methods are commonly known on the islands by the local communities and are intrinsically linked to psychiatry (Allen, 2010). It is important to note that the many different groups of people living on the islands ascribe themselves in varying degrees to traditional methods of healing and (sometimes at the same time) to (Western) biomedical methods. The two methods are not mutually exclusive in seeking care (Punski-Hoogervorst et al, 2021), yet an integrative approach combining the methods to optimise mental health care is still missing on the islands (Lynch, 2021). However, an integrative approach can also favour Western knowledge above traditional knowledge if there is no conscious effort to dismantle the power dynamics at play. Deliberately and thoughtfully integrating the healing practices that already exist within a community is a way of centring the worldviews of local communities so that they recognise themselves in health care communication.

Challenges to mental health care communication

Effective and genuine communication between individuals and health care providers is vital when discussing mental health issues. Explaining moods, thoughts, integrative aspects of behaviour, and possible treatment methods is impossible when there is doubt and ineffectual transmission of information (Satcher, 2001). The disparity in public access and level of health of BIPOC communities has been well documented (see example from the United States in Fiscella and Sanders, 2016). While on Bonaire, Saba, and Sint Eustatius access to allopathic mental health care is not limited by individual financial constraints, a barrier is perceived in the access and effectiveness of mental health care.

As underlined in the Post-Disaster Needs Assessment of Bonaire, there is a high demand for organised mental health treatment (World Bank, 2021). Health care providers observe that clients reach out to mental health care as a last resort in an often desperate situation. BIPOC communities are initially deterred from accessing mental health care due to mistrust, fear of treatment or discrimination, and differences in culture, language, and communication with the (often Dutch) providers. This mistrust can historically be traced to the criminalisation of mental illness and unethical experimental 'health care' practices performed on BIPOC communities (Gary, 2005; Vergès, 2020). For the ABCSSS islands in particular, the mistrust is rooted in colonial and postcolonial histories (Allen, 2010; Blom, 2015; Ansano, 2019). On the islands, mental health is strongly linked to spiritual well-being and religious belief systems. Traditional medicine and healing, as practised by the ancestors of current BIPOC communities, has been persecuted, considered invalid, and stigmatised by lawmakers and dominant religions and still has a complicated relationship with (Western) conventional medicine (Lynch, 2021). This history, along with the (universal) stigma on mental health, forms a significant barrier for mental health care communication and access to adequate and timely health care. The local communities' alienation from the predominantly Dutch and Dutch-speaking health care system has been reported in Faraclas et al (2022).

Additionally, ethnic disparities exist in how clients perceive their mental health care providers' cultural competence (Eken et al, 2021). Culturally competent care acknowledges and incorporates culture, cross-cultural relations, and 'vigilance towards the dynamics that result from cultural differences, and the adaptation of services to meet culturally-unique needs' (Cross, 1989). Eken et al (2021) elaborate by stating that adults from BIPOC communities 'were more likely to value seeing providers who shared or understood their culture'. These findings highlight the role of inclusion in mental health care and its communication and emphasise the importance of ensuring that mental health care providers have the competences to provide quality care. On the ABCSSS islands, a disparity can exist between the local BIPOC clients and their mental health providers, when (a large part of) the chief practitioners are, for example, White, from the European Netherlands, and lacking cultural competences. Further exacerbating this problem is the lack of opportunities for local training in mental health care professions. In a landmark collaboration in 2021, the ABCSSS islands' mental health institutions pledged to develop educational opportunities in mental health for local employees (Koninkrijk.nu, 2021).

E-health and language in mental health care communication

At Mental Health Caribbean (MHC) in particular, there is increasing attention paid to the cultural balance of the organisation and how this affects the quality of care. Besides investing more in strategy and recruitment for a personnel base that reflects the community it serves, MHC is developing culturally adapted e-health modules. This development in mental health care communication is a first for the ABCSSS islands. E-health is the use of the internet and digital resources alongside traditional face-to-face therapy to support treatment. Online, the patient can read about their treatment method, prepare for the next session, and chat with their therapist. E-Health has been a proven method of treatment, including in non-Western countries (Fu et al, 2020). Previous attempts at incorporating (Dutch) e-health in private mental health care practices in Curaçao have led to the exclusion of local and BIPOC communities. To reach island communities, MHC endeavours to translate the Dutch e-health modules into Papiamentu, the local language. This goes beyond literal translation and also involves sociocultural adaptations of texts and videos to appropriately reach the target audiences, such as visualising textual information, including storytelling, and changing examples to reflect Caribbean societies. This process is led by local therapists, content editors, and translators from the BIPOC communities that the modules are developed for. In this way, the project recentres Caribbean epistemologies 'through native language and dialects as a mode of decolonising' (Stewart, 2020, p 27). In Bonaire, Papiamentu has a long history of being neglected in education in favour of Dutch, but efforts to foreground Papiamentu in the education system have shown positive effects in school engagement and results (Beukenboom, 2021). Such effects are also expected when using Papiamentu instead of Dutch in mental health care communication and e-health in Bonaire.

Transformations in science communication practices and community-based engagement

The challenges of delivering inclusive and decolonised science communication in the ABCSSS islands have been illustrated in the previous sections with case studies from nature conservation and mental health care. The following sections build on the concepts introduced in this chapter and offer transformative practices in science communication that address the identified challenges. As this chapter shows, researchers, funders, and communicators can reflect on a number of practices in their work: (1) investing, supporting, and facilitating research and communication that is Caribbean-led; (2) recognising local knowledge and building long-term reciprocal collaborations; (3) reflecting on the dynamics of decision-making and implementing multi-vocality and co-creation in science communication; and (4) asking difficult questions and, in response, sometimes refusing to participate in research (Tuck and Yang, 2014). These practices can transform research and communication practices and build a more embedded and community-based engagement with research in the ABCSSS islands.

Caribbean-led research, nature conservation, and mental health care communication

On April 24, 2014, the Dutch Ministry of Education, Culture, and Science established the Caribbean Netherlands Science Institute, a facility that provides accommodation and infrastructure for researchers and students in the marine sciences. Following discussions on the role of this institute for the so-called 'Dutch Caribbean communities', the Dutch Research Council commissioned a report on the sustainable strengthening of knowledge systems in the Caribbean, in which the interviewed stakeholders reflected on the need to 'bring science closer to society' (Bijker and Wuite, 2021, p 6). The report emphasises that the organisation 'can only succeed if the people and institutions of those islands can claim ownership'. In the previous sections, the challenge of involvement and ownership arose repeatedly across examples in research and communication of nature conservation and mental health care. So, what is required to establish local ownership?

Ownership is fostered when people are involved as equitable partners from the beginning, and not (as in the case of the national parks) as an afterthought. The lack of access to opportunities for local researchers and mental health care professionals and their dependence on foreign universities and funding shown in this chapter are tackled by initiatives such as the PISKABON fisheries cooperative, the training for local health care professionals, and the hiring of Caribbean professionals. Local leadership in health care, research, and communication is critical. For example, island-based researchers from the ABCSSS islands initiated CaribResearch, a research foundation that aims to contribute to the resilience, progress, and sovereignty of their own communities. According to CaribResearch, it is the privilege and responsibility of local experts to initiate, realise, coordinate, and interpret local research. Led by local academics, the organisation has prepared a research agenda for the ABCSSS islands (CaribResearch Research Agenda, 2022). This is an important first step and an opportunity to include public engagement with research as one of the pillars of the research agenda for the islands.

As much as possible, Caribbean research projects and funding institutions should also involve researchers from and on the islands in paid leadership, research, and communication positions.

Recognising local knowledge and building long-term reciprocal collaborations

Caribbean institutions of higher education, health care, and research have the challenge to provide leadership in care, research, and education where traditional and local knowledge are just as valued.

Researchers, both local and non-local, are primarily trained in Western research methodologies that are not adapted for and do not take island contexts into account. As Walter Mignolo writes: 'We must confront the reality that our modes of questioning and even the answers they provide, often continue to be modelled after Western ways of thinking and interpretation' (cited in Stewart, 2019, p 66). This requires extra efforts on the part of researchers and health care institutions, both from and outside the Caribbean. By incorporating Indigenous and local methodologies and ways of knowing and communicating into research projects, local and non-local researchers can include communities in more equitable ways and 'broaden the imaginary of who can make a claim on science communication' (Orthia, 2020). For research and communication, this also means recognising oral traditions and local knowledge, appropriately valuing and remunerating knowledge providers, and practising humility when building long-term, reciprocal collaborations. Challenges like the lack of communication in Papiamentu and the inaccessibility of scientific results can be addressed by a tailored approach through long-term collaborations with experienced partners. There is an important role for local contributors as equal partners throughout the research process: not only in communication projects and products but also in project design, data collection, analysis, and academic publications. Caribbean research projects should also create space for all researchers to critically reflect on their own background, position, and training and how this influences their work (Trisos et al, 2021).

Reflecting on dynamics of decision-making and co-creation in science communication

To restructure unequal power relationships, researchers can employ multi-vocality in decision-making with regard to research and public engagement with science. Thus, pitfalls such as the profound disconnect between mental health communication and BIPOC communities, 'helicopter science' practices, or the lack of local impact of nature conservation research can be avoided. Engagement and ownership are also embedded through long-term co-creation strategies for public engagement (Sankatsing Nava and Hofman, 2018), as is the case in the Pelican conservation citizen science programme that led to the Important Bird and Biodiversity Area, or in Brenchie's Lab community maker space citizen science initiatives.

It is critical to reflect on the dynamics of decision-making in these collaborative projects: who has the power to decide? Whose timeline do we follow? Transparency in sharing agendas and goals is also vital to foster reciprocal collaborations. Well-written community engagement statements on proposals can successfully get research projects funded. However, it quickly becomes clear to the islanders when community consultation and engagement are genuine and when their sole purpose is a tokenistic 'box-ticking' exercise, getting pre-existing plans passed, or simply looking good (a public relations exercise). This then means that, when engagement is the goal, science communicators, researchers, and funders must plan and budget for genuine community engagement before designing projects (and calls for proposals) and build in flexibility and resources for multi-vocality, community wishes, needs, and subsequent project changes as a result of these collaborations.

Refusing research and asking the difficult questions

In her seminal book *Decolonising Methodologies*, Linda Tuhiwai Smith points to pertinent questions that can also be asked about research and science communication in the ABCSSS islands: who is the research for? Who are the owners of the research? Who will carry it out? Who will disseminate it? (Smith, 2021, p 10). From this chapter, additional questions can be added: who will design the public engagement plans? Who will benefit, but also, who defines what those benefits are? And finally, what are the conditions for this research to take place? Caribbean communities should be able to consider 'refusal' to participate in research as a viable option (Tuck and Yang, 2014). In 2020, Xiomara Balentina expressed one possible condition for participating in research. After a consultation session for a new research project, she wrote that 'keepers of traditional knowledge should only participate in research initiated by researchers from the Western university if the encounter results in decolonised spaces – Spaces in which the curriculum is reflective of different, yet equal voices from different geographical places' (Balentina, 2020). Communities of the ABCSSS islands can take Balentina's lead and formulate their own conditions for collaboration with foreign researchers.

Conclusion: Looking towards the future, imagining Caribbean ways to foster and exchange knowledge outside the academy

In the ABCSSS islands, science communication about nature conservation and mental health is often done by researchers and organisations from Europe and North America. With a number of notable exceptions, communication of mental health care and of nature conservation in the Caribbean has been a one-way street. These legacies shaped the exclusionary practices that have had a long-term impact on public engagement with science on the islands, in which both *science and science communication are done to us instead of done with us*. Therefore, a critical step in 'decolonising' science communication on the islands is to write 'our stories by us, for us, and for the world, rather than having stories written about us' (Stewart, 2019, p 17). In order to interrupt the existing relationship between science and society on the islands, there is an imperative to develop individual and institutional capacities for the decolonised production of knowledge together with and in service of the island communities. However, this also entails asking whether it is possible to decolonise science communication at all. This requires the critical interrogation of the role of science communication in strengthening scientific research and perpetuating its underlying structures. Changing how science is communicated does not necessarily change the nature of science itself, and recognising that 'easy absorption, adoption, and transposing of decolonisation is yet another form of settler appropriation' (Tuck and Yang, 2012; Hlabangane, 2018) is a crucial aspect of this process. Instead, Caribbean islanders should continue to take the lead in imagining and investing in alternative spaces and ways of fostering and sharing knowledge outside of the academy.

Engaged researchers in the Caribbean are moving away from the 'deficit model' that assumes a lack of knowledge about science that needs to be rectified. For science communication to be effective, it requires reframing not only from a monologue to a dialogue (Horst et al, 2017) but into a practice that fosters collaborative spaces where participants become co-researchers (Stewart, 2019, p ix). This is not easy to achieve within the current academic structures. In fact, it is not meant to be easy: these practices will unsettle our work and restructure not only science communication but also the underlying research practices. As Caribbean researchers and science communicators, we should continuously ask ourselves how we can build a research and communication practice that is more grounded locally, is embedded in questions relevant to islanders, and equitably involves stakeholders outside academia. The end-goal is to no longer have to communicate research back to non-academic island communities but instead to imagine, develop, and implement research and public engagement together with local communities. These profound changes can lead to more inclusive and locally contextualised scientific research and communication, whereby Caribbean communities play an active role in agenda-setting, formulating ethical frameworks, and designing and leading community-based public engagement with research in Aruba, Bonaire, Curaçao, Saba, Sint Maarten, and Sint Eustatius.

Acknowledgements

We would like to thank the following people for the conversations, reflections, and valuable contributions to this chapter: Antonio Carmona

Báez, Xiomara Balentina, Lysanne Charles, Alison Fischer, Corinne Hofman, Rosalba Icaza Garza, Joseph Sony Jean, Durwin Lynch, Luc Alofs, Stacey Mac Donald, Jurio Maduro, Eric Mijts, Emma de Mooij, Gert Oostindie, Pedro Russo, Taariq Ali Sheik, and the 'Under Construction' participants at the Royal Institute for South East Asian and Caribbean Studies. This chapter is part of the ongoing PhD research of the first author about public engagement with research in the Caribbean.

References

Allen, R.M. (2010) 'Hende a hasi malu p'e: popular beliefs in Curaçaoan culture', in N. Faraclas, R. Severing, C. Weijer, and L. Echteld (eds) *Crossing Shifting Boundaries: Language and Changing Political Status in Aruba, Bonaire and Curaçao*, Willemstad, Curaçao: Fundashon pa Planifikashon di Idioma, University of the Netherlands Antilles, pp 221–9.

Allen, R.M. (2018) 'Negotiating gender, citizenship and nationhood through universal adult suffrage in Curaçao', *Caribbean Review of Gender Studies*, 12: 299–318.

Ansano, R. (2019) 'Advances and challenges in safeguarding traditional medicine in Curaçao', in E. Falk (ed) *Traditional Medicine: Sharing Experiences from the Field*, np: ICHCAP, pp 106–15, Available from: https://ichlinks.geunnam.com/archive/materials/publicationsV.do?nation=KR&page=&ichDataUid=13829988369390600390

Balentina, X. (2020) 'The Western university in "exotic" spaces', *The Daily Herald*, 9 March.

Beukenboom, E. (2021) 'Het Papiaments treft geen blaam', *Antilliaans Dagblad*, 10 November.

Bijker, W. and Wuite, J. (2021) 'Dutch Caribbean research platform: towards the sustainable strengthening of the knowledge system in the Caribbean part of the Kingdom of the Netherlands', NWO, Available from: https://www.rijksoverheid.nl/documenten/kamerstukken/2021/06/16/eng-report-dutch-caribbean-research-platform

Blair, M.E. (2019) 'Toward more equitable and inclusive spaces for primatology and primate conservation', *International Journal of Primatology*, 40: 462–4. doi: https://doi.org/10.1007/s10764-019-00093-y

Blom, J.D., Poulina, I.T., van Gellecum, T.L., and Hoek, H.W. (2015) 'Traditional healing practices originating in Aruba, Bonaire, and Curaçao: a review of the literature on psychiatry and Brua', *Transcultural Psychiatry*, 52(6): 840–60.

Burns, M. and Medvecky, F. (2018) 'The disengaged in science communication: how not to count audiences and publics', *Public Understanding of Science*, 27(2): 118–30. doi: https://doi.org/10.1177/0963662516678351

Cannon, S.E. (2019) 'Decolonizing conservation: a reading list', Zenodo. doi: https://doi.org/10.5281/zenodo.4429220

CaribResearch Research Agenda 2022–26 (2022) 'CaribResearch: re-imaging our realities; strengthening our capacity to cope with crises', Foundation for Social Research of the Dutch Caribbean and the Region, Aruba, Bonaire, Curaçao, Sint Maarten, Saba and Statia.

Chilisa, B. (2019) *Indigenous Research Methodologies*, Thousand Oaks, CA: Sage Publications.

Connell, R.J. (2017) 'The political ecology of Maroon autonomy: land, resource extraction and political change in 21st century Jamaica and Suriname', PhD dissertation, UC Berkeley.

Cross, T.L. (1989) 'Towards a culturally competent system of care: a monograph on effective services for minority children who are severely emotionally disturbed', CASSP Technical Assistance Center, Georgetown University Child Development Center, ERIC Number: ED330171.

Dawson, E. (2014) '"Not designed for us": How science museums and science centres socially exclude low-income, minority ethnic groups', *Science Education*, 98(6): 981–1008.

De Santo, E.M., Jones, P.J., and Miller, A.M.M. (2011) 'Fortress conservation at sea: a commentary on the Chagos marine protected area', *Marine Policy*, 35(2): 258–60.

Dutch Caribbean Nature Alliance (2021) 'Annual report', DCNA. Available from: https://dcnanature.org/wp-content/uploads/2022/04/DCNA-YearReport-2021-Digital.pdf

Eken, H.N., Dee, E.C., Powers III, A.R., and Jordan, A. (2021) 'Racial and ethnic differences in perception of provider cultural competence among patients with depression and anxiety symptoms: a retrospective, population-based, cross-sectional analysis', *The Lancet Psychiatry*, 8(11): 957–68.

Faraclas, N., Kester, E.-P., and Mijts, E. (2022) '"We can do better than 'Chambuká'": toward inclusive language policy and practice in Bonaire', final report of the Maneho di Idioma research team, April.

Ferdinand, M. (2021) *Decolonial Ecology: Thinking from the Caribbean World*, Cambridge: Polity Press.

Fiscella, K. and Sanders, M.R. (2016) 'Racial and ethnic disparities in the quality of health care', *Annual Review of Public Health*, 37: 375–94.

Fu, Z., Burger, H., Arjadi, R., and Bockting, C.L. (2020) 'Effectiveness of digital psychological interventions for mental health problems in low-income and middle-income countries: a systematic review and meta-analysis', *Lancet Psychiatry*, 7(10): 851–64.

Gary, F.A. (2005) 'Stigma: barrier to mental health care among ethnic minorities', *Issues in Mental Health Nursing*, 26(10): 979–99.

Hall, M. and Tucker, H. (2004) *Tourism and Postcolonialism: Contested Discourses, Identities and Representations*, London: Routledge.

Hlabangane, N. (2018) 'Can a methodology subvert the logics of its principal? Decolonial meditations', *Perspectives on Science*, 26(6): 658–93.

Horst, M., Davies, S.R., and Irwin, A. (2017) 'Reframing science communication', in U. Felt (ed) *The Handbook of Science and Technology Studies*, Cambridge, MA: MIT Press, pp 881–907.

Koninkrijk.nu (2021) 'Samenwerking geestelijke gezondheidszorg tussen alle eilanden van het Koninkrijk', 30 October, Available from: https://koninkrijk.nu/2021/10/30/samenwerking-geestelijke-gezondheidszorg-tussen-alle-eilanden-van-het-koninkrijk/

Lynch, D. (2021) '"Embracing Brua": toward integrated mental health care in the Dutch Caribbean; opportunities and barriers for collaboration between traditional healers and biomedical practitioners', poster presentation, Dutch Caribbean Research Week, 14–18 June 2021. Funded with Dr Silvia de Groot Fonds/KITLV grant 2016.

Mabele, M., Sandroni, L.T., Collins, Y.A., and Rubis, J. (2021) 'What do we mean by decolonizing conservation? A response to Lanjouw 2021', 7 October, Available from: https://conviva-research.com/what-do-we-mean-by-decolonizing-conservation-a-response-to-lanjouw-2021/

Mac Donald, S. (2022) '"Life in paradise"? A social-psychological and anthropological study of nature conservation in the Caribbean Netherlands', PhD dissertation, Leiden University.

MacLeod, L. (2021) 'More than personal communication: templates for citing Indigenous elders and knowledge keepers', *KULA: Knowledge Creation, Dissemination, and Preservation Studies*, 5(1). doi: https://doi.org/10.18357/kula.135

Mbembe, A.J. (2016) 'Decolonizing the university: new directions', *Arts & Humanities in Higher Education*, 15(1): 29–45. doi: https://doi.org/10.1177/1474022215618513

Meredith, D. (2010) *Explaining Research: How to Reach Key Audiences to Advance Your Work*, Oxford: Oxford University Press.

Milieu Centraal (2022) 'Subsidies verduurzamen woning', 10 March, Available from: https://www.milieucentraal.nl/energie-besparen/energiesubsidies-en-leningen/subsidies-verduurzamen-woning/

Mijts, E., Ballantyne, J., and Rodriguez, C. (2022) UAUCU Student Research Exchange Collected Papers. Available from: https://www.ua.aw/wp-content/uploads/2022/05/UAUCU-BOOK-2022.pdf

NWO (2019) 'More funding available for Caribbean science', 5 February, Available from: https://www.nwo.nl/en/news/more-funding-available-caribbean-science

Olson, R. (2009) *Don't Be Such a Scientist: Talking Substance in An Age of Style*, Washington, DC: Island Press.

Orthia, L.A. (2020) 'Strategies for including communication of non-Western and Indigenous knowledges in science communication histories', *JCOM: Journal of Science Communication*, 19(2): A02. doi: https://doi.org/10.22323/2.19020202

Palmer, S.E. and Schibeci, R.A. (2014) 'What conceptions of science communication are espoused by science research funding bodies?', *Public Understanding of Science*, 23(5): 511–27.

Punski-Hoogervorst, J.L., Rhuggenaath, S.N., and Blom, J.D. (2021) 'Belief in Brua among psychiatric patients from Aruba, Bonaire, and Curaçao: results from an explorative study in the Netherlands', *Transcultural Psychiatry*, 59(3): 249–62.

Rai, N.D., Devy, M.S., Ganesh, T., Ganesan, R., Setty, S.R., Hiremath, A.J. et al (2021) 'Beyond fortress conservation: the long-term integration of natural and social science research for an inclusive conservation practice in India', *Biological Conservation*, 254: 108888.

Robbins, P, (2007) *Encyclopedia of Environment and Society*, Thousand Oaks, CA: Sage Publications.

Sankatsing Nava, T. and Hofman, C.L. (2018) 'Engaging Caribbean island communities with Indigenous heritage and archaeology research', *JCOM: Journal of Science Communication*, 17(4): CN06. doi: https://doi.org/10.22323/2.17040306

Satcher, D. (2001) 'Mental health: culture, race, and ethnicity; a supplement to mental health; a report of the surgeon general', US Department of Health and Human Services.

Sevold, T. (2020) 'How citizen science can help SIDS reach SDGs: experiences of open environmental research in Aruba', Available from: https://medium.com/@tony_87301/how-citizen-science-can-help-sids-reach-sdgs-8ef2920b7e00

Sint Maarten Nature Foundation (2011) 'Brown Pelican monitoring', Available from: https://naturefoundationsxm.org/research/monitoring/brown-pelican-monitoring/

Smith, L.T. (2021) *Decolonizing Methodologies: Research and Indigenous Peoples*, London: Zed Books.

Stewart, S. (ed) (2020) *Decolonizing Qualitative Approaches for and by the Caribbean*, Charlotte, NC: IAP.

Trisos, C.H., Auerbach, J., and Katti, M. (2021) 'Decoloniality and anti-oppressive practices for a more ethical ecology', *Nature Ecology & Evolution*, 5(9): 1205–12.

Tuck, E. and Yang, K.W. (2012) 'Decolonization is not a metaphor', *Decolonization: Indigeneity, Education & Society*, 1(1): 1–40.

Tuck, E. and Yang, K.W. (2014) 'R-words: refusing research', *Humanizing Research: Decolonizing Qualitative Inquiry with Youth and Communities*, 223–48.

Vergès, F. (2020) *The Wombs of Women*, Durham, NC: Duke University Press.

World Bank (2021) 'COVID-19 post-disaster needs assessment: Bonaire socio-economic assessment report', World Bank Group: Urban, Disaster Risk Management, Resilience & Land, Available from: https://bonairegov.com/fileadmin/user_upload/WorldBank-PDNA-SI.pdf

Zaitchik, A. (2018) 'How conservation became colonialism: Indigenous people, not environmentalists, are the key to protecting the world's most precious ecosystems', *Foreign Policy*, Available from: https://foreignpolicy.com/2018/07/16/how-conservation-became-colonialism-environment-indigenous-people-ecuador-mining/

Zanotti, L., Carothers, C., Apok, C.A., Huang, S., Coleman, J., and Ambrozek, C., (2020) 'Political ecology and decolonial research: co-production with the Iñupiat in Utqiaġvik', *Journal of Political Ecology*, 27(1): 43–66.

PART IV

The Globally Diverse History of Science Communication: Deconstructing Notions of Science Communication as a Modern Western Enterprise

13

Shen Kua's *Meng Hsi Pi T'an* (*c* 1095 CE): China's First Notebook Encyclopaedia as a Science Communication Text

Ruoyu Duan, Biaowen Huang, and Lindy A. Orthia

Introduction

The value for history, science, and culture of Shen Kua's iconic 11th-century text *Meng Hsi Pi T'an* 梦溪笔谈 (Brush Talks from Dream Brook) has been explored by modern researchers since the 20th century. As the earliest notebook encyclopaedia to be produced in ancient China, *Meng Hsi Pi T'an* caught the interest of mid-century science historians when philologist Hu Tao-Ching produced several acclaimed versions of the work, including a simplified version for lay readers (Hu, 1957; Wang and Zhao, 2011). Science history pioneers Joseph Needham and Wang Ling described *Meng Hsi Pi T'an* as 'a landmark in the history of science in China' and included English translations of parts of it in their own influential text *Science and Civilisation in China* (Needham and Ling, 1954, p 135; 1959). Shen has sometimes been labelled 'China's greatest scientist' (Holzman, 1958, p 260) and is relatively well known in China. However, the significance of his work is still under-appreciated elsewhere in the world, and perhaps in China too (Zuo, 2010).

In particular, *Meng Hsi Pi T'an* has generally been overlooked as a historical example of science communication. Even within China, only rarely have researchers characterised it as a popular science text or in other ways that might bring it under science communication's purview (for example Zeng and Guo, 2013). Zhang (2013, pp 366–7) is the notable exception: she argues that the book makes contributions to 'democratizing scientific knowledge', 'preserving grassroots science', and 'communicating science and technology to the lay public', considering it part of 'technical communication's enduring tradition'. In line with Zhang's view and the evidence she presents, this chapter makes the case for Shen Kua and his *Meng Hsi Pi T'an* to be incorporated into histories of science communication.

The significance of global histories of science communication

Published histories of science communication tend to focus on the West, particularly Britain and France, in the past 300 years (Orthia, 2020). This tends to be the case whether those histories are produced in Western or non-Western countries. They generally do not include communication cultures older than the so-called Scientific Revolution of the 16th and 17th centuries or cultures beyond Western Europe and its settler colonies, except when discussing the spread of Western science within them. Communication about knowledge held by other cultures – cultures that number in the thousands, span millennia, and hail from all over the world's six inhabited continents and innumerable islands – have largely been excluded from these histories.

This narrow focus has sometimes been justified as necessary to avoid both anachronism and inappropriate universalisation across cultures. Nonetheless, the narrowness of science communication histories contributes to a culture of racism and exclusivity within science communication research and practice (Orthia, 2020; Rasekoala and Orthia, 2020; Orthia et al, 2021). Science communication histories should – albeit cautiously – extend their reach far beyond the recent West to redress those consequences and cultivate a more culturally inclusive ethos within science communication.

As a first step, it must be acknowledged that the bodies of knowledge Western science is built upon were developed over centuries and millennia by people and cultures across the Afro-Asian landmass before being exported to Europe. More recently, Western science has also co-opted or incorporated knowledges from elsewhere in the world – the Americas, Australasia, the Pacific – that people and cultures from those places developed over centuries and millennia (Hountondji, 1997; Hobson, 2004, 2015; Harding, 2011). Because of that continuity, it is Eurocentric and racist to draw a boundary around 'scientific knowledge' and to place the recent West's knowledge inside it and everything else outside.

The word 'science' (and by extension, 'science communication') is policed by its Western proponents in a way that few other words are. Related words such as 'engineering', 'mathematics', 'knowledge', and 'communication' are not treated the same way in cross-cultural contexts. Like it or not, the label 'science' carries specific rhetorical power and prestige (Rochberg, 2010; Dear, 2012), so restricting its use may reinforce a Eurocentric and racist hierarchy of knowledge, with Western science at the top (Giglioni,

2007; Orthia, 2020). Because of this, many proponents of non-Western knowledge systems are reclaiming the 'science' label for their systems. For example, First Nations scientists and science, technology, engineering, and mathematics (STEM) professionals from the continent often known as Australia are claiming the word 'science' for Aboriginal and Torres Strait Islander knowledges, as 'Indigenous science' or similar (for example, Ball, 2015; Moggridge, 2018; de Napoli, 2018; Diamond, 2019; Noon, 2020).

This reclamation of the word carries the risk that Western science will continue to be seen as a benchmark for evaluating the worth of other knowledge systems, so some knowledge keepers and scholars prefer not to label their culture's knowledge as 'science' for this and other reasons. However, many advocates believe the strategic benefits of using the label outweigh this risk, and that the West's monopoly on the term must be broken. As a result of their actions, West-based science historians are rethinking their emphasis on labels and increasingly recognising diverse knowledge systems as part of the history of science (for example, Rochberg, 2016; Delbourgo, 2019), while respecting every knowledge system's specific sociocultural conventions, principles, and contents.

Similarly, space must be opened up for the communication practices surrounding those diverse bodies of knowledge to be recognised within the histories that science communicators write about their discipline and profession, if that's what knowledge keepers want for their cultural heritage. The specific label 'science communication' may have originated in the recent West, but as a discipline and realm of professional practice, science communication is at present global. Pretending that its scope is only Western is naïve at best and profoundly racist and exclusionary at worst. Much is to be gained from understanding its histories to be much older and more culturally diverse than most current histories suggest. What is needed is detailed historical studies of the world's thousands of cultures, exploring how people communicated about their knowledge of the world, with whom, in what circumstances, to what ends, and so forth, to build up a rich and diverse historical picture of global knowledge communication practices that might be categorised as 'science communication'.

As noted, representatives of some cultures do not want their knowledge categorised as 'science' or their knowledge communication to be categorised as 'science communication' for a range of reasons, including that those terms are not culturally appropriate. Accordingly, it is best if members of the cultures being studied lead historical studies in this space. Outsiders have done some important work to move this field along and can sometimes provide useful cross-cultural viewpoints in collaborative projects. But it is only from within a culture that a researcher or communicator can determine the culture's ideal orientation to global science communication and can deconstruct any false and inappropriate boundaries between the two.

Conversely, since representatives of many cultures *do* want their knowledge to be understood as science (or as equivalent to science), it follows that many want their culture's knowledge communication practices to be recognised as science communication too. But it is up to them to determine how this occurs. It is not for Westerners to gate-keep nor to imperialistically claim all knowledge communication under (Western) science communication's flag. This means global science communication discourse must let go of its attachment to Western models and open up space for redefining science communication in radically inclusive ways, ideally devised by diverse members of the world's many cultures.

Bringing *Meng Hsi Pi T'an* into science communication histories

This chapter is one response to that urgent need. It is the product of collaboration between two Chinese researchers – a postgraduate science communication student (Duan) and a communications academic (Huang) – and an Australian researcher who previously published on science communication history (Orthia). China has richly documented histories of science and of communication practices respectively, but existing histories of Chinese science communication that bring the two together mostly focus on the dissemination of Western scientific culture in China (for example Xiao, 2004; Yin and Li, 2020). Few attend to traditions for communicating about Chinese science and recognise them as examples of science communication.

In writing about an 11th-century Chinese text, the authors hope to shift the thinking of science communication teachers, researchers, and practitioners across the world to recognise what is similar between then and now and 'here' and 'there', as well as what is different. Indeed, by studying global science communication practices, all people may gain new perspectives on science communication in their own culture, which can only enhance practice, teaching, and research. The authors also hope this example (and the others in this book) will bolster the confidence of non-Western science communicators and students to more confidently introduce and highlight the historical achievements and contributions of their culture and what it has to offer global society today. More specifically, the example of Shen Kua and his *Meng Hsi Pi T'an* may prove useful for Chinese science communicators who wish to bring Chinese science history more squarely into their professional activities.

The chapter's discussion of this example begins by talking about Shen himself before moving on to a discussion of *Meng Hsi Pi T'an* and the Northern Sung context in which it was produced. It focuses specifically on elements relevant to science communication; however, it is noted that the level of commonality with current Western science communication norms should never be the benchmark for including any culture's knowledge communication practices in science communication histories (if knowledge keepers want them included). As such, we explore *Meng Hsi Pi T'an*'s communication contexts on their own terms as well as highlighting some resonances with current Western/global science communication practices in the hope that they will be of interest to readers.

A Northern Sung science communicator

To better understand *Meng Hsi Pi T'an*, it is first important to know its author. In 1090 CE, Shen Kua (沈括, 1031–95) retired to the Meng Hsi (Dream Brook) garden estate on the Yangtze River in what is now Zhenjiang. When he retired, he was a twice disgraced bureaucrat, but during his life Shen worked in diverse roles throughout China, including intellectual and management positions in three libraries, the Imperial Institute of History and the Bureau of Astronomical Observation, and administrative, planning, legal, and envoy roles for the state (Sivin, 1995; Zuo, 2010; Wang and Zhao, 2011). He acquired a wealth of knowledge and ideas from books he read, people he met, and professional problems he solved in a pattern of lifelong learning that began in childhood, educated by his mother and travelling with his father (Sivin, 1995; Zhang, 2013). This learning experience not only enriched his knowledge but also built a base from which he could communicate that knowledge in different styles to engage a broad range of readers. The travelling life came to an end after he was demoted to a rural area, and Shen quit rather than accept the unfavourable post (Wang and Zhao, 2011). During his last years, he enjoyed a quiet time at Meng Hsi and wrote down things he had learned, heard, and thought (Wang and Zhao, 2011). One of the products of that contemplative time was *Meng Hsi Pi T'an* (Shen, 2011), named for the writing brush and ink slab method Shen used to write it and the home where he wrote it.

As noted earlier, Shen has been labelled a scientist by recent historians, but given what he is most known for he might rather be considered an inspirational science communicator. The correspondences between Shen's work and science communication today include, most obviously, the fact that he wrote about many matters 21st-century people might associate with science or technology. Whatever label might be put on him, he was certainly a prolific communicator of science-related topics.

Shen's primary legacy was the compilation, communication, and critique of knowledge and ideas that circulated during his life. *Meng Hsi Pi T'an* contains mixed content, in the form of over 600 'jottings' grouped into 26 themes in most versions, including the version on which this study is based (Shen, 2011). The jottings' contents originate from four sources: published works, Shen's own research and thinking, non-elite people's knowledge and experience, and cultural debates about issues in science, politics, and philosophy. As Zuo (2010, p 256) puts it: 'He represents a singular moment in Chinese intellectual history, when "knowhow" practical thinking based on critical experiential inquiries attached itself to the rock of Confucian classical scholarship. Shen combined these heretofore contesting intellectual impulses, producing a coherent epistemology.'

Like science communication scholars today who increasingly recognise many sources of expertise including personal and professional experience, Shen did not merely admire knowledge established and advocated by well-known scientists or knowledge producers. Rather, he critically distinguished between different types of expertise and bodies of knowledge. For example, in jotting 314 (numbering from Shen, 2011) he asserted that book-based learning was an insufficient basis for medical expertise. He challenged medical books as being full of fallacies and promoted the value of professional experience learned in the field. He asserted that doctors should know about this problem and respond accordingly.

This overarching philosophy of scepticism about books framed the jottings within the 'Traditional Chinese Medicine' theme. Most of them reproduce the medical information found in books and attribute it to specific texts, but they also critique or correct that information when Shen believed the books were wrong. The context for this critical attitude was growing urbanisation and education in his society, and the displacement of people's traditional home remedies by elite medical practitioners who were often relatively unlearned (Sivin, 1995). In other words, new technologies, changing paradigms, and uncertainties about scientific expertise prompted Shen to intervene, just as similar circumstances in recent decades prompted the growth of fields such as critical science communication in the Western and global academy.

In this context, Shen presents as an informed communicator who not only knew the knowledge canon but also brought it into critical conversation with the lived experiential knowledge of diverse ordinary people. Shen argued that skills, technologies, and other things in the world were not only produced by elites but also created by workers, farmers, and other ordinary people (Li, 1974). In Zuo's words, he 'unconsciously [set] literati and artisans on a par', with more than one commoner given full credit for their contribution to Shen's book (Zuo, 2010, p 268). Or as Zhang (2013, p 366) puts it, Shen 'wanted to write down anecdotes, small talks, and folk tales that had been circulating in the back streets' in what was 'a bold departure from the conventional, official science writing of his time'. Zhang (2013, p 370) goes further, writing: 'Through conversations, observations, and experiments, [Shen] allowed the grassroots public to contribute their knowledge and talents … and to construct a public understanding of science together.'

The experience-based experts Shen cited included himself, as someone who frequently dealt with ill health. This multi-vocal (and auto-ethnographic?) approach was unusual for his time, in which writers usually simply copied their content from esteemed texts (Sivin, 1995) and 'called for the restoration of the perfect ancient order' (Zuo, 2010, p 257). Despite the millennium that separates them, there are echoes of similarity between *Meng Hsi Pi T'an* and pluralistic, reflexive, and critical approaches to science communication in the 21st century that credit experiential expertise.

Meng Hsi Pi T'an as a science communication text

Shen did make some original contributions to scientific knowledge in *Meng Hsi Pi T'an*, for example to mathematics, astronomy, and geology (Needham and Ling, 1959; Sivin, 1995; Huang, 2014). But it may be more appropriate to characterise Shen as a polymath, scholarly amateur who mostly sought to document, curate, and critique a broad range of information rather than primarily producing new knowledge as such. As Sivin (1995, p 11) puts it, 'Shen was writing for gentlemen of universal curiosity and humanistic temperament'. Some scholars have classified *Meng Hsi Pi T'an* as popular science because of its broadly accessible linguistic features such as direct and plain writing (Zeng and Guo, 2013; Zhang, 2013).

Even though Shen is recognised as the sole author of *Meng Hsi Pi T'an*, his extensive engagement with other people's ideas renders his work encyclopaedic in style. It is also encyclopaedic in vision, his jottings being relevant to a range of subjects and disciplines from music to humour to statecraft. These include many jottings that mathematicians, engineers, astronomers, taxonomists, geologists, medical scientists, geographers, and other STEM professionals (including science communicators) might take interest in today. However, Shen did not classify his jottings according to those disciplines (Pan, 2008; Kim, 2010; Zuo, 2010). For example, his discussion of steel-making techniques is classified under the theme 'Philological Criticism' (jotting 56), because it describes the correct names of iron products people commonly mistake for steel, and methods for making true (correctly named) steel. Similarly, he discussed eclipses (131, 139) and other celestial events under 'Chinese Numerology', an autopsy technique (209) under 'Administrative Affairs', an invention for suppressing toxic gas in a well (224) under 'Wisdom in Emergencies', and methods for calculating polyhedron volumes (301) under 'Crafts'. Other jottings, including one about petroleum (421), were left to 'Miscellanies', which Sivin (1995, p 35) suggests was for phenomena Shen could not fathom, and for which the challenge of finding their 'place in the cosmic schema … remains to be met by someone else'.

It might be deduced that themes within this classification scheme are based not on similar subject matter but on similar aims, audiences, and problem types. 'Wisdom in Emergencies' is clearly a theme developed for particular circumstances, along the lines of what might today be classified as risk management and risk communication. 'Administrative Affairs' was aimed at a government-bureaucrat audience, being a sort of policy communiqué theme. Zuo (2010) makes a strong case for the coherence of the category 'Jiyi' 技艺 ('Particular Skills', which in the translation used here is rendered 'Crafts'), whose jottings are unified by being step-by-step instructions for completing diverse practical tasks in order to achieve specific kinds of results. One typical example is movable type woodblock printing technology (307). Shen clearly recorded how workers print different Chinese characters and organise them into pages via movable woodblocks. Zuo (2010) speculates that, to be able to write this jotting, Shen must have watched the technology's inventor while he worked (the inventor was commoner Bi Sheng毕昇, *c* 970–1051). Zuo also proposed that Shen regarded himself as a practitioner training to use the apparatus, given how he described it (Zuo, 2018). His writing style engages readers by allowing them to quickly understand how to use the technology, because the description paints a dynamic picture of key procedural steps. There is resonance between this and what science communicators today might call technical writing, aimed at artisans and others working with their hands. Shen thus did the work of framing knowledge in a way that is fit for audience and purpose.

Throughout the book, Shen employed extensive use of figurative language including metaphors to explain abstract ideas by comparing them to everyday items (Zhang, 2013). However, for some concepts, metaphors were constitutive of their multiple meanings for both material understanding and learning philosophical life lessons. For example, in jotting 44 he described the properties of a concave mirror: it reflects close objects right side up but far away objects upside-down. He compared this to rowing a boat, in which pulling oars towards you will send the oar paddles away from you. He explained that this happens (to mirrors and oars) because there is 'a block in the middle' that pivots the image or the oar. He reflected that people are often blocked in life too, mistaking right for wrong or materiality for feeling, and they need to remove the block to see clearly.

Northern Sung society as a wellspring for science communication

Shen lived during the Northern Sung, a period of history characterised by a huge economy, a massive trade network that spanned the supercontinent of Afro-Asia (Hobson, 2004), and what Hobson (2004) has termed the first industrial revolution. Under the latter, China annually produced massive amounts of products and materials such as iron and paper, which were used to create diverse everyday objects for all classes of people, and it engaged in industrial-scale agriculture, resulting in large food surpluses (Hobson, 2004). These developments, plus a growing population and military endeavours, demanded new innovations in energy generation, planning, building, mapping, and other technologies. The Northern Sung was, in many ways, a mass, commercial society just as most countries are in the 21st century.

So was there something like a science communication movement in Sung China, linking knowledge, politics, publics, and new media? Certainly, there were changes to communication in the Northern Sung period. Education was broadened and made cheaper on the back of printing technology developments, and the state capitalised on this to promote selected ancient and current texts to the people (Sivin, 1995; Su, 2008). Examination culture during the Sung was accompanied by a more open dissemination of previously secret specialist knowledges, including in many areas the Western academy would associate with science (Kim, 2010). These developments broke the monopolies on information sources that had been managed and shared within higher classes in earlier dynasties and promoted book-based learning among all classes through increased accessibility (Su, 2008; Zhou, 2009). There was a large demand for books, which were published in both public and private sectors, distributed through markets and official libraries, and disseminated widely, including to rural and lower-class people, despite relatively small print runs (Nakayama, 1984; Sivin, 1995; Qing, 2001; Hobson, 2004).

Some of these books communicated scientific and technical information, and authors did often interweave this knowledge with political and philosophical perspectives (or, in Zuo's view, biases) (Kim, 2010; Zuo, 2010, p 264; Zhang, 2013). Kim (2010, p 218) discusses Sung scholars who felt a need to justify their passion for mathematics with respect to Confucian norms, or who sought to reconcile the two perspectives on knowledge; this suggests the prominence of epistemological conflict and strategically devised approaches when promoting specialised scientific and technical knowledge. Some Sung texts used pictures and poetry to communicate technical information to elites within a moral framework (Bray, 2007; Kim, 2010); arguably a form of promoting ideas to non-specialist audiences using accessible communication techniques. There were of course many important and large differences between Sung China and the globalised science communication community today (some discussed by Kim, 2010), and the necessarily brief glimpse of this complex society in the present chapter risks over-generalising and thus perhaps making misleading comparisons. But further research into this place and time from a science communication perspective is surely warranted.

It is unknown who might have encountered *Meng Hsi Pi T'an* during those years, as evidence is scarce. But given the societal context, it could be argued that laypeople and non-specialists of different sorts were likely to have read it and to have contributed to popularising this book in their period. Certainly in the centuries following Shen's life, readership for technical subjects broadened (Kim, 2010). Although some groups such as farmers and artisans were not generally literate, some laypeople still had opportunities to accept education provided by the academy of classical learning in the North Sung Dynasty (Ma, 2007). This kind of informal school offered classes to a broad range of people, whatever their identities and education backgrounds. When students studied at these schools, they not only learned how to read but to read different books, especially those printed by the government. This suggests some North Sung laypeople and non-specialists may have read Shen's book and broadly communicated its contents to others. Zhang (2013) contends that the practical orientation of the work and its user-friendly structure and style indicate its usefulness to the public, and that Shen deliberately sought to bring science to people excluded from formal education. She notes that court scientists at the time were not particularly committed to making science and technology more accessible, so Shen may have been exceptional in this regard.

Conclusion

Beyond readership during the Sung period, Shen and his text produced science communication impacts more recently. As noted in the introduction, Needham and Ling highlighted Shen in their important work that drew the attention of the Western academy and the Western public to Chinese science history in the 20th century. The fact that it is now not uncommon for scholars to talk about Shen as a scientist, and his work as scientific, is a testament to the usefulness of this person and text for increasing recognition of Eastern science in other parts of the world. That in itself is a form of science communication: a rhetorical act that provokes public and academic discourse about the meanings of 'science' in a cross-cultural context.

Recognising *Meng Hsi Pi T'an* within science communication histories can have a material impact on the discipline's culture of inclusion. The origin of this chapter was an Australian postgraduate class in science communication history in which one of the authors (Duan) was an international student from China. While the course focused on the recent West in line with the emphasis in the literature, she campaigned to focus her main research assignment on this text, enabling her to claim space within the discipline for her own culture and heritage. With the support of the class teacher (Orthia), and where appropriate the assistance of an auditing visitor (Huang), she was then able to apply the Shen case to practical science communication tasks in her degree programme, for example to consider how she might use this different perspective on science communication in her role as an Explainer at a science centre. Challenging received ideas of what science is enables science communicators to ask audiences to think about science broadly, for example to ask how they would define the knowledge in *Meng Hsi Pi T'an* given it was published before the word 'science' was used in English. This kind of example provides rhetorical tools to science centre Explainers and other science communication professionals, who can consequently be more persuasive and critical when they introduce aspects of non-Western scientific culture to diverse visitors.

Indeed, introducing this example to the class broadened the imagination of the teacher, the other students, and auditing visitors as to how science communicators might re-conceptualise science communication history to be more inclusive. This mission should be of fundamental importance to the discipline as the body of science communicators in the world grows, in China and most other countries around the globe (Schiele et al, 2012; Xu et al, 2015; Gascoigne et al, 2020). The authors hope science communicators around the world can reflexively reconsider their own knowledge and join together in making efforts to expand the picture of science communication history and other fields of science communication.

References

Ball, R. (2015) 'STEM the gap: science belongs to us mob too', *Australian Quarterly*, 86(1): 13–19.

Bray, F. (2007) 'Agricultural illustrations: blueprint or icon?', in F. Bray, V. Dorofeeva-Lichtmann, and G. Métailié (eds) *Graphics and Text in the Production of Technical Knowledge in China*, Leiden: Brill, pp 521–67.

De Napoli, K. (2018) 'Indigenous astronomy to revitalise the Australian curriculum', IndigenousX, 8 December, Available from: https://indigenousx.com.au/indigenous-astronomy-to-revitalise-the-australian-curriculum/

Dear, P. (2012) 'Science is dead; long live science', *Osiris*, 27(1): 37–55.

Delbourgo, J. (2019) 'The knowing world: a new global history of science', *History of Science*, 57(3): 373–99.

Diamond, C. (2019) 'Young Indigenous women invited to join STEM Academy', IndigenousX, 5 August, Available from: https://indigenousx.com.au/young-indigenous-women-invited-to-join-stem-academy/

Gascoigne, T., Schiele, B., Leach, J., Riedlinger, M., Lewenstein, B.V., Massarani, L. et al (eds) (2020) *Communicating Science: A Global Perspective*, Canberra: ANU Press.

Giglioni, G. (2007) 'Irritating Experiments: Haller's Concept and the European Controversy on Irritability and Sensibility, 1750–90 (review)', *Bulletin for the History of Medicine*, 81(3): 662–4.

Harding, S. (ed) (2011) *The Postcolonial Science and Technology Studies Reader*, Durham, NC: Duke University Press.

Hobson, J.M. (2004) *The Eastern Origins of Western Civilisation*, Cambridge: Cambridge University Press.

Hobson, J.M. (2015) 'The Eastern origins of the rise of the West and the "return" of Asia', *East Asia*, 32: 239–55.

Holzman, D. (1958) 'Shen Kua and his Meng-ch'i pi-t'an', *T'oung Pao*, 46(3/5): 260–92.

Hountondji, P.J. (ed) (1997) *Endogenous Knowledge: Research Trails*, Dakar, Senegal: Codesria.

Hu, T.-C. (ed) (1957) *Meng Hsi Pi T'an Xin Jiao Zheng*, China: Chung Hwa.

Huang, H. (2014) 'Zhu Xi's research on the natural science of *Meng Hsi Pi T'an*', *Guizhou Social Sciences*, 292(4): 28–32.

Kim, Y.S. (2010) 'Confucian scholars and specialized scientific and technical knowledge in traditional China, 1000–1700: a preliminary overview', *East Asian Science, Technology and Society*, 4: 207–28.

Li, Q. (1974) 'Shen Kua and his Meng-ch'i Pi-t'an', *Journal of Integrative Plant Biology*, 3: 10–13.

Ma, H. (2007) 'A study of the academy in the North Song Dynasty and analysis of its characteristics', *Chinese Local History Records*, 10: 26–32.

Moggridge, B. (2018) 'From STEM to stern', *The Walkley Magazine*, 10 January, Available from: https://medium.com/the-walkley-magazine/from-stem-to-stern-db33bc1e3897

Nakayama, S. (1984) *Academic and Scientific Traditions in China, Japan, and the West*, translated by J. Dusenbery, Tokyo: University of Tokyo Press.

Needham, J. and Ling, W. (1954) *Science and Civilisation in China, Volume I*, Cambridge: Cambridge University Press.

Needham, J. and Ling, W. (1959) *Science and Civilisation in China, Volume III*, Cambridge: Cambridge University Press.

Noon, K. (2020) 'Indigenous science can save us', IndigenousX, 23 January, Available from: https://indigenousx.com.au/indigenous-science-can-save-us/

Orthia, L.A. (2020) 'Strategies for including communication of non-Western and Indigenous knowledges in science communication histories', *JCOM: Journal of Science Communication*, 19(2): A02.

Orthia, L., Hikuroa, D.C.H., Nabavi, E., Rochberg, F., and de Vos, P. (2021) '3 reasons to study science communication beyond the West', The Conversation, 12 January, Available from: https://theconversation.com/3-reasons-to-study-science-communication-beyond-the-west-152237

Pan, T.-H. (2008) 'A study of classification of *Meng-ch'i Pi-t'an*'s items', *Journal of Zhen Jiang College*, 21(4): 1–6.

Qing, Y. (2001) 'The development of media technology and the reform of publishing in Song', *Journal of Zhejiang University (Humanities and Social Sciences)*, 31(5): 157–9.

Rasekoala, E. and Orthia, L.A. (2020) 'Anti-racist science communication starts with recognising its globally diverse historical footprint', LSE Impact of Social Sciences Blog, 1 July, Available from: https://blogs.lse.ac.uk/impactofsocialsciences/2020/07/01/anti-racist-science-communication-starts-with-recognising-its-globally-diverse-historical-footprint/

Rochberg, F. (2010) 'Beyond binarism in Babylon', *Interdisciplinary Science Reviews*, 35(3–4): 253–65.

Rochberg, F. (2016) *Before Nature: Cuneiform Knowledge and the History of Science*, Chicago, IL: University of Chicago Press.

Schiele, B., Claessens, M., and Shi, S. (eds) (2012) *Science Communication in the World: Practices, Theories and Trends*, Dordrecht: Springer.

Shen, K. (2011) *Brush Talks from Dream Brook*, translated by H. Wang and Z. Zhao, Chengdu: Sichuan People's Publishing House and Paths International.

Sivin, N. (1995) *Science in Ancient China: Researches and Reflections*, Aldershot: Variorum.

Su, Y.-Q. (2008) *Northern Sung Dynasty Books and Ancient Chinese Movement*, Zhejiang: Zhejiang University Press.

Wang, H. and Zhao, Z. (2011) 'Translator's preface', in H. Wang and Z. Zhao (eds) *Brush Talks from Dream Brook*, Chengdu: Sichuan People's Publishing House and Paths International, pp 1–14.

Xiao, X. (2004) 'The 1923 scientistic campaign and Dao-discourse: a cross-cultural study of the rhetoric of science', *Quarterly Journal of Speech*, 90(4): 469–92.

Xu, L., Huang, B., and Wu, G. (2015) 'Mapping science communication scholarship in China: content analysis on breadth, depth and agenda of published research', *Public Understanding of Science*, 24(8): 897–912.

Yin, L. and Li, H. (2020) 'China: science popularisation on the road forever', in T. Gascoigne, B. Schiele, J. Leach, M. Riedlinger, B.V. Lewenstein, L. Massarani et al (eds) *Communicating Science: A Global Perspective*, Canberra: ANU Press, pp 205–26.

Zeng, J.-P. and Guo, L. (2013) 'The strategy of science popularization in new media era', *Modern Communication (Journal of Communication University of China)*, 1: 115–17.

Zhang, Y. (2013) 'Examining scientific and technical writing strategies in the 11th century Chinese science book *Brush Talks from Dream Brook*', *Journal of Technical Writing and Communication*, 43(4): 365–80.

Zhou, B.-R. (2009) 'The arrival of the era of printing and the study style of the Sung Dynasty society', *He Nan Social Sciences*, 17(3): 125–7.

Zuo, Y. (2010) 'The production of written knowledge under the rubric of *Jiyi*', *East Asian Science, Technology and Society*, 4: 255–73.

Zuo, Y. (2018) *Shen Gua's Empiricism*, Cambridge, MA: Harvard University Asia Center.

14

Making Knowledge Visible: Artisans, Craftsmen, Printmakers, and the Knowledge Sharing Practices of 19th-Century Bengal

Siddharth Kankaria, Anwesha Chakraborty, and Argha Manna

Introduction

Existing research in the science communication literature has commented on how science and technology are seen predominantly through the prism of Western/Eurocentric knowledge paradigms (Neeley et al, 2020). Dawson (2019) highlights how this Eurocentrism spills over to sites and instances of science communication, where minority communities often feel marginalised and their knowledge underrepresented. This is also the case in the context of India, where the history of 'modern' scientific and technological enterprises is a story often retold through the lens of European colonisation of the Indian people and the subsequent transfer of Western knowledge paradigms within the Indian subcontinent (Phalkey, 2013; Chakraborty et al, 2020). Phalkey (2013) further notes that historical accounts about scientific practices in India have not paid enough attention to such practices situated within Indian society, which could be rectified through 'the study of institutional, social, political, economic and cultural contexts with a focus on the experiences of practitioners so that a practice-oriented understanding of science in India can emerge' (Chakraborty et al, 2020, p 371).

In an attempt to address these lacunae, this chapter presents a historical account of printmaking practices in colonial Bengal (predominantly

19th century), while also describing the messy entanglements of science, arts, craftsmanship, local technologies, and society in that era. Using existing literature and archival material, it highlights the following: (a) that coloniser–colonised relationships cannot be viewed through prisms of predefined binaries and need to further account for pre-existing sociopolitical divisions within the Indian subcontinent; (b) how caste-based divisions of labour prevalent in colonial Bengal historically influenced the complex intersection of different knowledge systems; and (c) there is an urgent need for the inclusion of various forms of knowledge systems in societies with strong exclusionary practices and for moving towards a more holistic understanding of knowledge-making and sharing practices. In doing so, this chapter places the experience of the local printmaking communities of colonial Bengal at the centre and argues that the role of marginalised groups such as the 'lower' caste communities was crucial to the rise and development of the Bengali printing and knowledge-sharing practices.

The chapter begins with a brief description of the context of the sociocultural demographics of colonial Bengal (the capital of British-occupied India) and its knowledge-sharing ecosystems comprising educational institutions, printing establishments, and a variety of local knowledge production and brokering practices. It then delves deeper into the politics of knowledge-making and sharing and explains how existing sociocultural divisions like caste and class served as crucial factors in determining what constituted knowledge in those times. The chapter specifically foregrounds dichotomies like 'theory versus practice', 'clean versus dirty', and 'intellectual versus manual' in terms of the different forms of knowledge existing at that time, and how these interacted with the nexus of power, privilege, and politics in colonial Bengal. It then goes on to highlight the role of local actors and Indigenous knowledge systems in enabling the rise of the printing press as a technology in colonial Bengal. Building on these contexts, it uses the case study of Battala publications to demonstrate how technological transfers in printmaking heavily relied on the expertise and 'practical' knowledge of the 'lower' caste communities. It further reflects on how the contributions of 'lower' caste communities played a critical role in upscaling and democratising the printing practices of colonial Bengal, thereby making the transition from orality to literacy in Bengali possible and eventually allowing the written word to become accessible to people outside dominant caste groups. The chapter concludes by discussing some of these findings in the light of present-day knowledge-sharing (science communication) practices and proposing relevant learnings for making contemporary science communication more inclusive, reflexive, and decolonised.

Colonial Bengal: sociocultural contexts and knowledge-sharing ecosystems

In the 19th century, most of the princely states and regions of modern-day India, Pakistan, and Bangladesh (hereafter referred to as India for convenience) were under British colonial rule. There was a stark sociocultural divide not just along the axis of the British colonisers and the colonised Indians but also within Indian society itself based on class and caste privileges. This divide was palpable in the spatial locations occupied by various sections of the population. The Europeans, 'upper' caste elites, and 'lower' caste communities occupied different areas of the city of Calcutta that signified their relative 'stations' in society (Ray, 1978; Sinha, 1978). The caste system was built on the foundations of discrimination, oppression, and social stigmatisation of 'lower' caste communities that were expected to be devoted to 'serving' the White British colonisers and the 'upper' caste elites of colonial India (Dutta, 1981; Banerjee, 1989). 'Upper' caste families, especially the Hindu Brahmin ('upper' caste) and Abhijata (aristocratic) families, controlled discourses and epistemologies of knowledge that were heavily influenced by Western conceptions of knowledge and science (Bandyopadhyay, 1980). The city's 'lower' caste populace comprised working-class people such as farmers, artisans, manual labourers, domestic servants, and so on that were often nomadic and were constantly in flux between Calcutta and its neighbouring villages – which they originally hailed from in search of jobs, work opportunities, and a better life (Leach and Mukherjee, 1970).

But around the mid-19th century, a new category of people emerged: the middle-class Bhadralok (gentlemen) that consisted of a diversity of service people ranging from shopkeepers, small merchants, landlords, clerks, administrative officers, and officials working at the East India Company. The Bhadralok classes also embodied the changes happening in society at that time, with new traditions of knowledge entering their worldview and shaping them alongside their traditional ones. The rise of the Bhadralok was further fuelled by a parallel rise of knowledge-sharing institutions in the early 19th century – which included schools, colleges, book societies as well as printing presses – all of which served as conduits of cross-cultural exchanges between the colonisers and the colonised. The opposite construct of the Bhadralok was the Chotolok, a pejorative term literally meaning smaller or lesser people, often deeply rooted in the oppressive Hindu caste system. The term has historically been used as a way of diminishing and marginalising people lacking sociocultural capital (Sinha and Bhattacharya, 1969).

Before discussing the impact of such sociopolitical divisions on the knowledge-sharing practices of colonial Bengal, it is important to take a brief look at the development of some of the main knowledge-sharing institutions of 19th-century Bengal. In 1800, the British government set up the Fort

William College for the purpose of teaching newly recruited British civil servants and officers the local languages and cultures of India (Roebuck, 1819; Raj, 1986). In retrospect, the college also played an important role in the standardisation of the Bengali language and its printing (Khan, 1976, pp 397–426). For instance, the college employed many Sanskrit-educated Bengali teachers who wrote textbooks on the Bengali language and helped standardise its grammar, orthography, and literary forms (Ghosh, 2006, pp 66–106). Unfortunately, most of these standardisations were heavily biased towards 'upper' caste Brahmanical and elitist Western conceptions of knowledge. Another important centre of learning, the Hindu College, opened in Calcutta in 1816 and served as a privileged centre of European learning 'reserved exclusively for sons of respectable Hindoo families' (Raj, 2007, p 160) that imparted Eurocentric pedagogical instruction (Kerr, 1852; Raj, 1986, p 160). In the same year, the Calcutta School Book Society was also set up as a joint venture between Indians working at the Fort William and Hindu Colleges that helped develop, commission, and purchase secular (non-religious) textbooks – in English as well as local languages – for imparting relevant scientific knowledge within its affiliated colleges (Khan, 1976, pp 200–346; Raj, 2007).

These educational institutions were supported by the parallel rise of printing presses in Bengal. In the years preceding 1800, Bengal saw the opening of 12 printing presses, which mostly published in English and catered to the needs of fresh British recruits arriving in India at that time (Khan, 1976, pp 397–426). Established by Europeans in 1800, the Serampore Mission Press, which is also referenced later in this chapter, was different in that it played a seminal role in the introduction and development of the Indigenous printing enterprises of colonial Bengal and in promoting Western arts, philosophy, and religion in forms and languages that were accessible to the local populations of India (Raj, 1986, p 117; Raj, 2007, p 172).

The Serampore Mission Press was one of the first large-scale, structured printing enterprises and was well known for its quality of work, service, and a diversity of technological innovations in printing in Indigenous languages. Members of the press were also responsible for helping standardise the grammar, literary forms, and orthography of the Bengali language, which was until then predominantly an oral language (Marshman, 1859; Bingham, 1951, pp 101–2). Over time, the press expanded its printing activities to more than 40 Indian languages and was staffed by about 150 employees, many of whom were locals, and thus allowed for significant cross-cultural synergies to emerge subsequently. The Serampore Mission Press also played a crucial role in the rise of Indian journalism and was responsible for printing some of the first Indian-language newspapers, magazines, and pamphlets, including the Bengali weekly *Samachar Darpan* (News Mirror), the Bengali science magazine *Digdarshan* (loosely translated as 'Showing Directions'), and the English monthly periodical, *The Friend of India* (which was later incorporated into *The Statesman*, a leading daily newspaper still in circulation in India) (Bingham, 1951, pp 100–8; Khan, 1976, pp 200–346).

Hegemony and marginalisation: knowledge politics in colonial Bengal

It is important to foreground at this stage the differences in what was actually considered knowledge by the 'upper' caste versus the 'lower' caste communities of colonial Bengal. The 'upper' castes and the emerging Bhadralok classes were keen on occupying prominent positions within newly emerging urban structures such as offices, businesses, judiciary, schools, and other jobs brought about by the colonial administration, all of which made education and learning increasingly valuable for them (Raj, 1986, p 116). The Bhadralok were thus constantly adapting and re-purposing various forms of European learning as a means of consolidating their social position as powerful intermediaries between the *colonising* British officials and the *colonised* Indigenous populations (Raj, 2007, p 164). For them, a formal education in European knowledge systems served as a gateway to their sociopolitical power in colonial Bengal, as was also retrospectively reflected in the Simon Commission Report in 1930: 'The school is the one gate to the society of the Bhadralok' (Simon and Indian Statutory Commission, 1930, p 24; Raj, 1986, p 116).

On the one hand, the Bhadralok classes prioritised certain 'higher' forms of knowledge as more desirable, such as linguistics, mathematics, astronomy, astrology, law, grammar, literature, and music, all of which had historically been pursued in India for mastering religious scriptures such as the *Vedas* or cultivated through extensive ceremonies and rituals. Most of these 'higher' forms of knowledge were also communicated in Sanskrit (historically considered the 'language of the cultivated' or the 'erudite Brahmins' in India) or sometimes in a more ornate version of Bengali (which was significantly different from the Bengali spoken by the masses in that era), which automatically granted the Bhadralok a systemic monopoly over other knowledge forms (Raj, 1986, pp 109–13).

On the other hand, knowledge systems pursued by the 'lower' castes were considered relatively 'lesser' forms of knowledge and mostly comprised practical, vocational, and skill-based knowledge systems. These forms of knowledge were often the product of the 'lower' caste communities' extensive (but often enforced) engagement with a diversity of skill-based manual labour ranging from carpentry, pottery, spinning, weaving, metallurgy, farming, and as will be further elaborated in this chapter, various forms of printmaking, craftwork, and artisanal practices. Most of these vocational knowledge systems were highly niche and had predominantly been orally transmitted across generations using local, vernacular languages. The lack of textual accounts, the need for physical tools for mastering these knowledge forms, and the relatively long periods of intergenerational interactions and skill-based training needed for transferring such knowledge-sharing practices all collectively restricted the scale, geographical reach, and efficacy of these 'lower' caste communities' knowledge transfer activities (Raj, 1986, pp 109–13).

In his essay 'Hermeneutics and cross-cultural communication in science', Kapil Raj employs the term 'clean knowledge' (which did not require the 'soiling of one's hands') to depict a fundamental ideal of the 'upper' caste elites and contrasts it with the 'practical *savoir-faire*' knowledge of the 'lower' caste communities (which was often considered 'soiled' or 'dirty'), thereby describing a 'hierarchical stratification' of knowledge systems in colonial Bengal (Raj, 1986, pp 109–13). This social stratification of knowledge also extended to the types of Western scientific knowledge being prioritised and pursued by the Bhadralok classes in 19th-century colonial Bengal. For instance, subjects like 'mathematics, algebra, Euclidean geometry, astronomy, Newton's laws of motion, hydrostatics, mechanics, optics and pneumatics' that were mostly mathematical and theoretical in nature were seen as a natural extension of their ideas of 'clean knowledge' and were highly sought after by the Bhadralok scions, as opposed to more experimental, hands-on, and vocational sciences that were considered more 'dirty' and 'messy' (Raj, 1986, p 119).[1] In light of these contexts, we argue that 'lower' caste communities were *dually marginalised* and were subjected to two modes of marginalisation: one via the colonial domination that affected the entire Indian population and the other via domination by 'upper' caste Hindus that specifically impacted the 'lower' caste communities.

Indigenous printmaking practices: local actors and the rise of Battala publications

Based on the history of printing in colonial Bengal narrated through a Eurocentric perspective, it is easy to conclude that the development of Bengali printing as well as its orthography and typography were primarily a result of the contributions of the various Christian missionaries and printing presses set up by the British government in the 19th century. But, it is important to remember that their motivation for developing Bengali printing was predominantly borne out of religious, administrative, and political agendas of spreading Christian ideals, 'ruling' over the local populations, and understanding these Indigenous cultures with the intention of better exploiting them (Khan, 1976, pp 200–346). In light of this contextualisation and in an effort to move towards a more decolonised history of Indian printmaking, it is imperative to shift the focus away from these institutional efforts that have already been well documented elsewhere (Khan, 1976, pp 200–346; Gupta, 2014, pp 57–67) and instead foreground the relatively lesser-known contributions of local printmakers, craftsmen, and artisans in the development of Bengali printing.

While the contributions of Nathaniel Brassey Halhed in compiling a grammar of the Bengali language and that of Charles Wilkins in helping build a printing press capable of printing in Indian languages are well known (Khan, 1976, pp 397–426), several local actors such as Panchanan Karmakar and the people trained by him were also critical in the development of a font of Bengali types that ultimately ushered in the era of Bengali printing (Chakravorty and Chaudhuri, 2009, p 8). Briefly, Wilkins had helped set up the Serampore Mission Press and was particularly skilled in the craft of printmaking, having himself cut the punches for many of the Bengali font types.[2] During his time at Serampore, he employed and personally trained Panchanan Karmakar, a particularly talented Indian blacksmith and craftsman by profession (and caste), hailing from the village of Tribeni (Khan, 1976, p 272). Wilkins taught Karmakar the nuances of type-casting, punch cutting, and printing press operations; and with surprisingly little help from any of the European printing institutions, they together developed font types from scratch for not just Bengali but many other Indian languages. Of particular note is their font of Persian types in Nasta'liq characters and another font of Persian types in Naskh characters (Reed, 1887, p 70; Khan, 1976, pp 200–346). At the Serampore Mission Press, Karmakar was also responsible for developing a font of Devanagari types that was required by William Carey for printing various works of Sanskrit grammar and scholarly literature. This was a monumental task requiring more than 700 separate punches in order to accurately represent the large number of compound letters used in the Devanagari script. To achieve this, Karmakar soon appointed an assistant called Manohar, who belonged to the same community and caste as him. Manohar later went on to become Karmakar's son-in-law and helped him develop a more refined font for Bengali type. Over time, Karmakar also transmitted this niche *savoir-faire* and vocational knowledge to many other local craftsmen within his community, including his grandson, and together their collective contributions went on to play a seminal role in the development, scaling, and refinement of printing in Indian languages (Khan, 1976, pp 200–346; Ghosh, 2006, pp 109–16; Bose, 2019, pp 10–67).

By the early 19th century, the Serampore Mission Press was printing in multiple Indian scripts including Devanagari (Hindi and Sanskrit), Arabic, Persian, Telugu, Punjabi, Marathi, Oriya, Kannada, and even non-Indian scripts such as Chinese, Burmese, Greek, and Hebrew, in addition to English. Even though the rise of the modern Indian printing enterprise was a result of close coordination and supplementation of skills, technical know-how, and lived expertise between European and Bengali individuals, the contributions of the native printmakers, artisans, and craftsmen have been disproportionately neglected and erased during the documentation of these historical practices, and therefore specifically merit foregrounding here (Khan, 1976, pp 19–54; Chakravorty and Chaudhuri, 2009).

The gradual rise in the printing of Indigenous texts and scripts at the beginning of the 19th century also enabled a massive boost in the use of illustrations, engravings, and visual elements within these printed publications over the next few decades (Sen, 1984). Rising demand for illustrated books further paved the way for English-imported engraved copper plates for printing more of these illustrations. Over time, local printmakers and craftsmen made considerable improvements in the quality and aesthetics of these copper-plate engravings (Gupta, 2014, pp 55–6; Bose, 2019, pp 10–67). In this regard, it is important to note that many of these copper-plate craftsmen and engravers were originally blacksmiths or goldsmiths (belonging to 'lower' caste communities), who were trying to branch out into new realms of secondary work sources by suitably adapting their traditional skill sets and culturally inherited knowledge (Paul, 2013, pp 2–18; Bose, 2019, pp 10–67). After the 1830s, the copper-plate engravings started to wane in popularity and were gradually replaced by woodcuts. There were a number of reasons for this: copper-plate engravings were much more expensive to design, required separate machinery for printing text and visuals, and often led to production delays since the copper plates had to be imported from England and could not be made locally from scratch. On the other hand, woodcuts were significantly cheaper to make and design, did not require separate machinery (given that the printing of text and images could be combined together), and could easily be manufactured locally. However, the shift to woodcuts did pose some challenges for the metal craftsmen and engravers, who, unlike earlier, found it more difficult to re-purpose their traditional metal engraving skills to wood-based mediums and took some time to adapt and excel in this new medium (Bose, 2019, pp 10–67).

By 1859, there were about 46 printing presses in operation in colonial Calcutta, many of them concentrated in the region around Chitpur Road called Battala (or Bartala) in the northern part of the city. The printing presses of the Battala area mostly used wooden presses, thus allowing them to produce a large number of inexpensive books, magazines, pamphlets, and other publications in bulk, often estimated to be around 15 million pages a year (Khan, 1976, p 420). Most of these publications were characterised by their own distinct style, iconography, and printing techniques and soon started to be recognised and metonymically referred to as Battala itself (Gupta, 2014, p 58).

In terms of its visual design, the Battala soon became well known for their vibrant use of illustrations, visual elements, and graphical motifs. Their distinct visual design was achieved manually using woodcut prints that often drew inspiration from the aesthetics, motifs, and techniques employed by Kalighat artists renowned for their *patas* or paper-based paintings on a number of religious and secular themes. These Kalighat painters often 'fused diverse strands of Indian art: illuminated manuscripts from Bengal under the Palas, Mughal, and Rajput schools of painting, British watercolour and shading' (Chakravorty and Chaudhuri, 2009, p 15). Similarly, the woodcuts often combined 'pictographic additive structure, flat decorative style, ornamentation, and narrative character' of these traditional art forms and seamlessly translated them into the 'simplified linear graphic forms' that later became characteristic of Battala woodcut prints (Sen, 1984; Bose, 2019, pp 10–67). Not surprisingly, developing such artistic woodcuts for printing Battala visuals often required coordination between various highly skilled workers including artists, engravers, printmakers, and printers (Sen, 1984; Bose, 2019, pp 10–67). Many of these features of Battala publications, especially in terms of their design, production, and dissemination, played an important role in making knowledge more accessible in colonial Bengal, as shall be described in more depth in the next section.

Democratising knowledge: role of Battala publications

The emergence of the genre of Battala publications and the emancipative ecosystem it engendered challenged the existing social and intellectual order. It further undermined the stronghold of the 'upper' caste Brahmin and Bhadralok classes who not only controlled the social, cultural, and political agendas of those times but also contributed to the prioritisation of certain forms of knowledge over others as a tool for upward social mobility. With the rise of Indigenous printing and the growth in the publication of cheap Bengali publications like Battala, the Brahmin–Bhadralok stronghold on knowledge production and dissemination was challenged in a number of ways.

The development of Bengali printing technologies, tools, and processes significantly lowered the entry barriers for starting a printing business, a phenomenon especially demonstrated in the case of the multiple Battala printing outlets. From the mid-19th century, many 'lower-' and middle-class Bengali entrepreneurs and businessmen – described as 'petty Bhadraloks' by Anindita Ghosh in her book *Power in Print* (2006) – began setting up their own printing businesses and subsequently employed a lot of local craftsmen, artists, and printmakers in this process. Furthermore, many of the Battala books were printed on cheap paper and bound in thin paper jackets that made them especially inexpensive and thereby affordable to the working classes. This growth spurt in cheap Battala publications also led to the diversification of books and types of publications, as well as the themes covered and the audiences catered to by these publications (Bhadra, 2011).

Such Battala publications covered a broad diversity of topics ranging from 'scandals, almanacs, farces, erotica, and romances, as well as biographies, histories, religious texts, and educational textbooks' and also produced Bengali translations of literature derived from other cultures (Sen, 1984; Banerjee, 1989, p 44; Chakravorty and Chaudhuri, 2009, p 13). These Battala publications, therefore, offered a healthy mix of 'entertainment, information and instruction' and became especially popular among the working classes, for whom it opened the doors to a rich diversity of knowledge for the first time (Banerjee, 2008, p 46). Moreover, in addition to catering more specifically to the tastes and interests of working-class audiences, Battala publications also directly addressed many social issues such as remarriage rights for widows, the eradication of infant marriages, female education and emancipation, modification of the caste system, and the regulation of societal drinking habits (Khan, 1976, pp 434–5).

The kind of Bengali language used in Battala publications was also especially lucid and accessible in terms of its vocabulary, grammar, orthography, and typography, and could thus be enjoyed by the majority of the 'lower' caste and working-class communities, who often did not have access to formal education via schools and colleges (Banerjee, 1989, pp 44–6). Another interesting feature of the increasing involvement of working-class individuals in the printing enterprise was that many of these entrepreneurs, artists, craftsmen, and printmakers had a much deeper understanding of the surrounding local cultures and market trends. Most of these cheap Battala books were liberally sold in public spaces including weekly markets, village fairs, train stations, street corners, and even via travelling street hawkers, all of which helped increase the reach of the publications to a greater diversity of audiences (Khan, 1976, pp 441–3). Employing these street hawkers was a particularly ingenious strategy, as they served as book salespersons going from door to door with a huge stack of books and magazines. These hawkers made it possible for people to browse through these publications at a leisurely pace – something that readers couldn't do at bookshops, depots, and railway stations nearly as conveniently – and helped bring many uninitiated audiences into the readership of Battala. Not surprisingly, these hawkers were much more successful at selling books than other physical outlets like bookshops, particularly in the case of female hawkers who managed to sell a large number of books to working-class women (Khan, 1976, pp 441–3).

On a similar note, artists and craftsmen involved in designing the illustrations, woodcuts, and engravings used in Battala publications were also able to frequently experiment with different styles, aesthetics, and techniques in these outputs. Given the rapidly evolving tastes of the working class and the concomitant growth of the printing industry, these artists and craftsmen contributed significantly to the dynamic and changing visual languages of Battala (Banerjee, 1989; Bose, 2019, pp 10–67). The constant dialogue with local tastes, interests, and other 'market forces' of their audiences also made Battala publications much more creative, reflexive, and secular in nature, so much so that the 'craftsman-printmakers had a reciprocal relationship with the market; their commercial works did not simply reflect the existing demands or dominant ideologies of the time, but also helped create them' (Bose, 2019, p 22).

Because of these features, Battala publications became especially accessible to the working classes and served as a notable enterprise that challenged the hitherto unquestioned gatekeeping of literary, educational, and knowledge systems by the 'upper' caste, Brahmin–Bhadralok classes. In her book, Ghosh attributes Battala literature as having served as a 'bridge between orality and literacy' within the popular culture of colonial Bengal, and whose rising popularity helped highlight the unfulfilled void in Bengali literary works that were specifically aimed at and enjoyed by the working-class populations of 19th-century Bengal (Ghosh, 2006; Banerjee, 2008, p 44). This excerpt from Sumanta Banerjee's review of Ghosh's book summarises this sentiment well:

> By giving printed form to this plurality of voices, ranging from the patois of the Calcutta streets to the slang of the village marketplace, from the domestic speech of women to the dialect of Muslim boatmen, Battala publications offered a counterculture vis-a-vis the 'high' literature of the educated bhadraloks, who mainly followed the hegemonistic model of a uniform 'standardised' Bengali written style. (Banerjee, 2008, p 44)

These contributions of 'lower' caste communities to printmaking thus allowed them to engage in knowledge-making and sharing practices that were relevant as well as accessible to their own communities and thereby provided them with a potential tool to break away from their cycle of marginalisation. The artisans and craftsmen of Battala, therefore, contributed to these knowledge ecosystems at two levels: first, by directly participating in the production of knowledge outputs and publications, as well as the printing technologies needed for creating them; and second, by creating and shaping locally relevant and accessible content for their own communities – both of which eventually helped democratise knowledge in colonial Bengal.

Conclusion: Centring the marginalised – lessons for contemporary science communication

In this chapter, we have described specific instances of the emergence of new knowledge being made possible due to the contributions of local and Indigenous actors, whose roles had hitherto been marginalised and sometimes even erased from mainstream accounts of history. For instance, the role of

Karmakar and his associates in the development of Bengali printing has received far less attention compared to Wilkins and his European colleagues. Further, 'lower' caste artisans skilled in specialised fields of knowledge such as Kalighat artforms, metallurgy, and printmaking coming together to produce a new style of copper-plate engravings, and subsequently, the genre of Battala publications using woodcut prints, are both great examples of how such marginalised communities helped challenge the hegemony of 'upper' caste communities as gatekeepers of knowledge. Such examples especially foreground the role of marginalised actors in the rise and development of the printing enterprise in colonial Bengal. It is also important to underline here that most available works describing the contributions of 'lower' caste and marginalised communities, sparse as they are, have been authored by those belonging to the 'upper' caste and socio-economically privileged communities.[3] This naturally has a bearing on how these texts understand, interpret, and document these marginalised histories and sharply highlights the urgent need for creating more spaces and opportunities for marginalised voices to emerge and share their own stories.

Importantly, the historical case studies and examples described in this chapter offer multiple learnings for contemporary science communication and public engagement practices. For instance, the case study of Battala publications succinctly highlights several benefits of enabling and inviting local actors and working-class communities to participate in the process of knowledge-making and sharing. Firstly, these local contributions were critical in ensuring that the content of Battala publications catered to their target audiences' tastes and interests and that both its language and illustrations remained particularly accessible to these working-class audiences. Secondly, the contributions of local actors such as the street hawkers greatly helped market and sell Battala publications in ways that allowed them to reach a greater diversity of (often uninitiated) audiences. Thirdly, the active involvement of members from these local and working-class communities also helped generate significant impact in terms of shaping social trends and spreading awareness about critical issues and eventually contributed to cultivating a stronger sense of social belonging, inclusion, and participation within these communities. These examples offer valuable insights for designing participatory knowledge-making and sharing practices that are better centred around local and Indigenous contexts; providing marginalised communities with a direct agency to shape, own, and control their narratives; and ensuring that these practices remain receptive, reflexive, and respectful to alternative forms of knowing, seeing, and doing.

While this chapter has extensively discussed the role of social identifiers like caste and class in determining who participated in these printmaking practices of colonial Bengal, it is also imperative to address the axis of gender here. The chapter has previously mentioned the contributions of women hawkers in selling Battala publications to a newly emerging women readership. In addition, several accounts of the history of Battala publications have also commented on the rise of specific 'women's literature' and the role of women readership in shaping its contents, including the societal navigation of gendered roles over time and the rise of more assertive female voices (Ghosh, 1998, 2002; Bannerjee, 2008). However, in the literature research that we undertook for writing this chapter, no instances of women directly participating in the workforce or practice of printmaking were found. A plausible reason for such an absence of women printmakers could be the predominantly (socially perceived) 'masculine' nature of the kinds of manual labour required for such printmaking jobs. It is also important to note that most of the 'lower' caste men engaged in these printmaking jobs were already excluded from the history of printmaking due to their social location, and if indeed there were women actively engaged in such printmaking processes, it is very likely that their contributions would have been made doubly invisible. To move beyond such calculated speculation, more detailed knowledge about the everyday lives of men and women belonging to the 'lower' caste communities of colonial Bengal is needed, and this is an area that could benefit immensely from further historical investigations.

A similar argument can also be made for considering the axis of race and ethnicity here. Historians of science in India have consistently challenged monolithic representations of science as a Western paradigm. Kapil Raj specifically talks about the location of constructing modern science as not being the Western laboratory but instead as the 'intercultural contact zones' (such as those found in 19th-century colonial Bengal), and in doing so, fundamentally 'relocates knowledge-making in different and infinitely more heterogeneous milieus' (Raj, 2007, p 223). As this chapter demonstrates, such 'intercultural contact zones' were not merely linear juxtapositions of the cultures of Europe and India, and therefore need to be viewed with much more granularity and nuance. This chapter demonstrates that (among other compounding factors) such historical narratives urgently need to better account for Indigenous social structures such as caste and class, and the inherent heterogeneity and dynamics they bring with them. Using historical accounts of the contributions of marginalised communities in colonial Bengal, this chapter foregrounds the importance of culture, context, and community in terms of studying the juxtaposition of knowledge and practice in everyday lives. It also highlights the need for studying both knowledge and practice in symbiosis and the use of this learning to build on current efforts for decolonising and diversifying contemporary science communication research and practice. Specifically, our chapter serves as a timely reminder of why the act of decolonising science and science communication cannot happen if scientific knowledge and technological practice stay in silos, especially with science being privileged at the expense of technology.

In foregrounding the contributions of marginalised actors such as 'lower' caste communities in the development of printing technologies and knowledge-sharing practices of 19th-century colonial Bengal, this chapter makes a critical contribution to the emergent discourses on inclusive science communication practices. Specifically, social inclusion in science communication has received a considerable amount of academic attention in recent times from the community of science communication researchers and practitioners alike (Dawson and Jensen, 2011; Falk et al, 2012; Massarani and Merzagora, 2014; Finlay et al, 2021; Matias et al, 2021; Orthia et al, 2021). Most of these conversations around inclusionary practices in science communication discuss the need for engaging with distant publics (Matias et al, 2021) such as marginalised communities (Dawson, 2014; Feinstein and Meshoulam, 2014; Streicher et al, 2014), creating specific engagement programmes that respond to their needs (Dawson, 2019) and developing ways to reach such marginalised communities in order to take the message of science and technology to them. However, in this chapter, we posit a completely different argument through these case studies, reminiscent of Orthia et al (2021), by centring the work of marginalised communities and acknowledging their role as producers of techno-scientific knowledge themselves. Instead of framing these communities merely as audiences of techno-scientific communication practices, they are mainstreamed here as fundamental contributors to the very process of techno-scientific knowledge-making and sharing, thereby placing their intrinsic but often forgotten contributions into sharp focus. The chapter illustrates that the picture of science and technology transfer and adoption is indeed varied, complex, and multidirectional when these histories are narrated from the perspectives of the marginalised, excluded, and oppressed communities very often relegated to the bottom rungs of the social ladder. These observations further highlight the need for multidisciplinary research that meaningfully incorporates people's lived experiences and sociocultural contexts in order to improve trans-cultural understandings of knowledge-making and sharing, develop critical tools for transformation, as well as build strategies for engaging in effective and ethical knowledge-making and sharing practices.

Importantly, the chapter also adopts a radically different interpretation of science communication and public engagement by more broadly seeing it as 'knowledge-making and sharing practices'. Here 'knowledge' can be seen to comprise topics not just within the natural sciences but also within the social sciences and the humanities as well as other vocational and skill-based knowledges. Further, these practices of 'making' and 'sharing' knowledge often regard 'science' and 'science communication' as a seamless continuum of each other, rather than as two distinct and discrete activities. Incorporating such an encompassing understanding of knowledge-making and sharing, which is both discipline-agnostic as well as more porous to acknowledging a diversity of expertise and lived experiences, serves as an essential step towards decolonising the understanding of both science and science communication. It furthermore questions and opens up the boundaries of what is valuable knowledge and meaningfully moves towards making these practices and processes more inclusive, diverse, and decolonised.

Notes

1 However, the Bhadralok classes' pursuit of medicine remains a notable exception and can perhaps be attributed to the sense of divinity associated with saving human lives (Raj, 1986, p 119).

2 Understanding the development of these font types merits a separate chapter in itself and interested readers are encouraged to read the works of Reed (1887), Marshman (1859), Khan (1976), and Gupta (2014) in this regard.

3 We write this with an acute awareness of our own privileged positionalities in undertaking this project. Nevertheless, we hope that this chapter will serve as a platform for more diverse and marginalised accounts of knowledge-making and sharing practices to build upon eventually.

References

Bandyopadhyay, B. (1980) *Sambad patre sekaler katha* (3rd edn), vol 1, Calcutta: Bangiya Sahitya Parisad.

Banerjee, S. (1989) *The Parlour and the Streets: Elite and Popular Culture in Nineteenth Century Calcutta*, Calcutta: South Asia Books.

Banerjee, S. (2008) 'The story of "Battala"', *Economic & Political Weekly*, 43(15): 44–6.

Bhadra, G. (2011) *Nyara Battalay Jay Kaw'bar*, Kolkata: Chhatim Books.

Bingham, G.E. (1951) 'The Baptist Mission Press of Calcutta', *Baptist Quarterly*, 14(3): 100–8.

Bose, A. (2019) 'Modernism and the graphic art of Bengal', PhD dissertation, Visva Bharti University, Santiniketan, Available from: http://hdl.handle.net/10603/301402

Chakravorty, S. and Chaudhuri, S. (2009) 'Printing and book production in Bengal: an exhibition at Rabindranath Tagore Centre', Jadavpur University, British Council and Indian Council for Cultural Relations, Kolkata, Available from: https://archive.org/stream/IndiaPrintingAndBookProductionInBengal/India-Printing%20and%20Book%20Production%20in%20Bengal%20%20_djvu.txt

Chakraborty, A., Raman, U., and Thirumal, P. (2020) 'Tracing science communication in independent India: towards an institutional and people's history', in T. Gascoigne, B. Schiele, J. Leach, M. Riedlinger, B.V. Lewenstein, L. Massarani et al (eds) *Communicating Science: A Global Perspective*, Canberra: ANU Press: pp 371–394.

Dawson, E. (2014) 'Reframing social exclusion from science communication: moving away from "barriers" towards a more complex perspective', *JCOM: Journal of Science Communication*, 13(2): 1–5.

Dawson, E. (2019) *Equity, Exclusion and Everyday Science Learning: The Experiences of Minoritised Groups*, London: Routledge.

Dawson, E. and Jensen, E. (2011) 'Towards a contextual turn in visitor studies: evaluating visitor segmentation and identity-related motivations', *Visitor Studies*, 14(2): 127–40.

Dutta, P. (1981) *Kolikatar Itibritta*, Calcutta: Pustak Bipani.

Finlay, S.M., Raman, S., Rasekoala, E., Mignan, V., Dawson, E., Neeley, L. et al (2021) 'From the margins to the mainstream: deconstructing science communication as a White, Western paradigm', *JCOM: Journal of Science Communication*, 20(1). doi: https://doi.org/10.22323/2.20010302

Ghosh, A. (1998) 'Cheap books, "bad" books: contesting print-cultures in colonial Bengal', *South Asia Research*, 18(2): 173–94. doi: 10.1177/026272809801800204

Ghosh, A. (2002) 'Revisiting the "Bengal Renaissance": literary Bengali and low-life print in colonial Calcutta', *Economic and Political Weekly*, 37(42): 4329–38.

Ghosh, A. (2006) *Power in Print: Popular Publishing and the Politics of Language and Culture in a Colonial Society, 1778–1905*, New Delhi: Oxford University Press.

Gupta, A. (2014) 'The Calcutta School-Book Society and the production of knowledge', *English Studies in Africa*, 57(1): 55–65. doi: https://www.tandfonline.com/doi/abs/10.1080/00138398.2014.916908

Falk, J., Osborne, J., Dierking, L., Dawson, E., Wenger, M., and Wong, B. (2012) 'Analysing the UK science education community: the contribution of informal providers', Wellcome Trust, UK, Available from: https://kclpure.kcl.ac.uk/portal/files/8448633/Falk_et_al_WT_review_2012.pdf

Feinstein, N.W. and Meshoulam, D. (2014) 'Science for what public? Addressing equity in American science museums and science centers', *Journal of Research in Science Teaching*, 51(3): 368–94.

Kerr, J. (1852) *A Review of Public Instruction in the Bengal Presidency, from 1835 to 1851*, Calcutta: Baptist Mission Press.

Khan, M.H. (1976) 'History of printing in Bengali characters up to 1866 (vol 1)', PhD dissertation, University of London, United Kingdom.

Leach, E. and Mukherjee, S.N. (eds) (1970) *Elites in South Asia*, vol 10, Cambridge: Cambridge University Press.

Marshman, J.C. (1859) *The Life and Times of Carey, Marshman, and Ward: Embracing the History of the Serampore Mission*, vol 2, London: Longman, Brown, Green, Longmans & Roberts.

Massarani, L. and Merzagora, M. (2014) 'Socially inclusive science communication', *JCOM: Journal of Science Communication*, 13(2), Available from: https://jcom.sissa.it/article/pubid/JCOM_1302_2014_C01/

Matias, A., Dias, A., Gonçalves, C., Vicente, P.N., and Mena, A.L. (2021) 'Science communication for social inclusion: exploring science & art approaches', *JCOM: Journal of Science Communication*, 20. doi: https://doi.org/10.22323/2.20020205

Neeley, L., Barker, E., Bayer, S.R., Maktoufi, R., Wu, K.J., and Zaringhalam, M. (2020) 'Linking scholarship and practice: narrative and identity in science', *Frontiers in Communication*, 5: 35.

Orthia, L.A., McKinnon, M., Viana, J.N., and Walker, G.J. (2021) 'Reorienting science communication towards communities', *JCOM: Journal of Science Communication*, 20(3). doi: https://doi.org/10.22323/2.20030212

Paul, A. (2013) *Unish Shatoker Kathkhodai Shilpi Priyogopal Das*, Calcutta: Signet Press.

Phalkey, J. (2013) 'Focus: science, history, and modern India; introduction', *Isis: An International Review Devoted to the History of Science and Its Cultural Influences*, 104(2): 330–6.

Raj, K. (1986) 'Hermeneutics and cross-cultural communication in science: the reception of Western scientific ideas in 19th-century India', *Revue de synthèse*, 107(1): 107–20.

Raj, K. (2007) 'Defusing diffusionism: the institutionalization of modern science education in early-nineteenth-century Bengal', in *Relocating Modern Science*, London: Palgrave Macmillan, pp 159–80.

Ray, A. (ed) (1978) *Calcutta Keepsake*, Calcutta: Riddhi-India.

Reed, T.B. (1887) *A History of the Old English Letter Foundries: With Notes, Historical and Bibliographical, on the Rise and Progress of English Typography*, London: Elliot Stock.

Roebuck, T. (1819) *The Annals of the College of Fort William from the Period of Its Foundation 1800 to the Present Time*, Calcutta: Pereira.

Sen, S. (1984) *Battalar Chhapa O Chhobi* (2nd edn), Calcutta: Ananda.

Simon, J.C. and Indian Statutory Commission (1930) 'Indian Statutory Commission report', vol 1, Survey.

Sinha, P. (1978) *Calcutta in Urban History*, Calcutta: Firma KLM.

Sinha, S. and Bhattacharya, R. (1969) 'Bhadralok and Chhotolok in a rural area of West Bengal', *Sociological Bulletin*, 18(1): 50–66.

Streicher, B., Unterleitner, K., and Schulze, H. (2014) 'Knowledge rooms: science communication in local, welcoming spaces to foster social inclusion', *JCOM: Journal of Science Communication*, 13(2): C03.

Conclusion: Advancing Globally Inclusive Science Communication – Bridging the North–South Divide through Decolonisation, Equity, and Mutual Learning

Elizabeth Rasekoala

Introduction

The chapters in this book have shown that science communication and public engagement practices, initiatives, and research take place in highly different contexts, scenarios, and settings around the globe. Yet, much of the discourse, practice, and research in the field is still predominantly about and conducted in the Global North. Together, these chapters have elaborated on what a truly globally representative, multidimensional, and inclusive dialogue on the state of the field of science communication and public engagement would entail. They have also highlighted the challenges, opportunities, and strengths in the scenarios, contexts, and settings that are underrepresented, excluded, and marginalised from the science communication 'mainstream', both in the Global North and the Global South.

These globally inclusive discourses are highly pertinent given systemic global inequalities and the differential science education and science communication systems in many parts of the globe. We need a broader framework for science communication that will transcend the Global North–South divide to explain the challenges in different contexts and proffer transformative solutions globally. This interconnected network enhanced by the mutual learning it facilitates across this Global North–South divide would further advance the transformation of science communication, rather than the limited 'globalised' framework predicated on Eurocentric dominance that has become normalised across the field. In this context, the imperatives of sociocultural and language plurality and diversity in the practices of science communication also come to the fore. In their interrogation of the presumed universality of science, African scholars Olukoshi and Nyamnjoh state that

> [i]t is of utmost importance to understand that science is not free of culture. It is, rather, not only full of culture but also does not function independently of its culturally-rooted and specific language bearing practitioners and their vested interests, whatever their claims to a lay status and neutral stance. (Olukoshi and Nyamnjoh, 2011, p 19)

This profound assertion can similarly be attributed to science communication as well.

There is still a long way to go before we can even begin to realise a critical level playing field of a truly globally inclusive craft and footprint of science communication – meaning science communication practices, methodologies, and reference points that are radically framed and informed by the very diverse indigeneity and cultural heritage of their localised populations and contexts. These disparate 'context-situated' transformative science communication frameworks can, in alignment, create a globally inclusive and level playing field through their very differentiated and yet locally responsive landscapes, liberated from the trappings of hegemonic practices and tendencies. The way forward lies in interrogating and further re-imagining how novel developments can be conceptualised, initiated, and then scaled-up to deliver the deep-rooted structural and systematic transformation of the science communication field itself. And how will we know when we are getting there? When diversity, equity, and inclusion become the principal indicators and metrics of the measurement of excellence as a global standard in science communication and public engagement practices, rather than an optional extra. This aligns with the truism of the adage 'what is not measured, and not counted, is never truly valued".

Bridging the Global North–South divide through decolonisation: contestations, shifting paradigms, and emancipative approaches

Eurocentric hegemony and ideology are reflected in the dominant narratives that attest to scientific advancement across the globe in discourses that seem to suggest that only people from the Global North have advanced the scientific enterprise while people of the Global South have made very little contribution. And yet, history shows us that this is not the case, with myriad examples of the contributions of ancient civilisations in Africa, the Middle East, Latin

America, and Asia, and their ground-breaking contributions to the advance of science across the world. Such historical exemplars have been highlighted by the chapters in Part IV of this collection, 'The Globally Diverse History of Science Communication: Deconstructing Notions of Science Communication as a Modern Western Enterprise'. Contemporary developments are highlighted in the chapters in Part II, 'Science Communication in the Global South: Leveraging Indigenous Knowledge, Cultural Emancipation, and Epistemic Renaissance for Innovative Transformation'. In this regard, it is also then very clear that the pseudo-historical memory that sustains Eurocentric hegemony and ideology needs to be challenged if the decolonisation of science communication is to be achieved.

In her seminal book *Half of a Yellow Sun*, the internationally acclaimed African writer Chimamanda Ngozi Adichie (2007) provides a vivid illustration of how challenging the decolonisation mind-set change can be even for the most avowedly 'liberal-minded' Europeans. In the early part of Adichie's book, there is a cringe-worthy social interaction between Richard, a White British journalist (who has purportedly come full of good intentions to tell the genuinely empowering stories of the Igbo people of the region, in the postcolonial era, as a means of redressing past omissions), and a Nigerian academic, Okeoma. During this interaction at a dinner, Richard, who has just arrived in the region, declares to Okeoma his longstanding fascination and interest in the ancient bronze Igbo-Ukwu artefacts that were found in that part of Nigeria. "I've been utterly fascinated by the bronzes since I first read about them", enthuses Richard. "The details are stunning. It's quite incredible that these people had perfected the complicated art of lost-wax casting during the time of the Viking raids. There is such marvellous complexity in the bronzes, just marvellous." To which Okeoma replies, "You sound surprised, as if you never imagined *these people* capable of such things".

This illuminating snapshot from Adichie's book is one that is all too wearily familiar to many practitioners from the Global South regions of the world from their professional and social interactions with their Global North counterparts. In this specific case, it directs us to use the bronze Igbo-Ukwu artefacts to initiate a discussion on the interrogation of the issue of values within the context of the subjective Eurocentric gaze as it views Africa's contributions to the historical, contemporary, and future advancements in the scientific enterprise. We are driven to ask questions of this Eurocentric gaze, such as what values, perspectives, and priorities drive the scientific enterprise in the Global South vis-a-vis the Global North, and why does this matter, in the context of the decolonisation of museums, of scientific knowledge, and of its inclusive communication? (See, for example, Howarth, 2018; Orthia, 2020; Finlay et al, 2021; Shoenberger, 2022.)

These interrogations should lead to further and deeper understandings given that science communication and public engagement with science as is understood by the Eurocentric mainstream is much more recent than older science communication traditions across many parts of the Global South, which predate it by centuries and illustrate the globally diverse historical trajectories of the field (Rasekoala and Orthia, 2020). These traditions and practices have been marginalised and disenfranchised by the colonial experience to the detriment of the cultural, linguistic, and socio-economic advancement of these nations, societies, and communities. The longstanding impacts of this, unfortunately, still persist to the present day, including their impacts on the advancement of the scientific enterprise, its innovations, and its communication (see Alatas, 2006; Maldonado-Torres, 2007; Gunaratne, 2009; Asante, 2011; Okere, 2011; Aman, 2018; Calvente et al, 2020; R'boul, 2021).

The legacy of the profound epistemic loss to Africa of the centuries-long extroversion of its scientific knowledge and innovation assets by European nations during the colonial era is one that still hinders the comprehensive liberation of its agency and autonomy of knowledge production and communication. In the words of the eminent African philosopher Paulin Hountondji, this 'in effect, ties it hand and foot to the apron strings of the West' (Hountondji, 1997, p 1). The American foreign journalist and academic Howard French (2021) articulates as his primary motivation for writing his landmark book *Born in Blackness: Africa, Africans and the Making of the Modern World, 1471 to the Second World War* the need to fill in crucial gaps and to overturn the assumptions, narratives, and myths that exclude Africans and Africa from the formation, innovations, and advances of the modern world. His book challenges the longstanding erasure of African innovation assets and the common narratives of the historical relationship between Europe and Africa, particularly the idea that European nations were somehow always superior to their African counterparts – whether in economic wealth, scientific knowledge, and innovation, state power, or technological prowess.

The decolonisation agenda in science communication needs to encompass the full gamut of the locations where Eurocentric hegemony, ideology, and pseudo-historical memory reside. It requires the decolonisation of mind-sets, perspectives, language, programmes, and exhibits. It needs to provide answers to profound questions, such as how do we re-calibrate the discourses, exemplars, and themes that infuse this globalised science communication arena? It needs to address the challenges inherent in scientific infrastructure-based programmes such as science centres and science museums, which provide the public with interactive, hands-on engagement and experiences with science and technology kits, games, gadgets, displays, and exhibitions; and the non-infrastructure-based expositions of science communication, such as the myriad science festivals, science fairs, olympiads, quizzes, public lectures, competitions, and other public engagement platforms such as those undertaken in partnership with science journalists and the mass media on radio, television, and social media.

Furthermore, the decolonisation of the science communication agenda needs to address the mechanisms of the pseudo-historical memory that characterise the situated discourses around the exhibitions of objects and artefacts taken from the Global South. These mechanisms of interrogating the true history of objects are important museum concepts that enable the capacity building of individuals and visitors to these museums, contributing to the shaping of historical and collective memory. At the same time, they also mitigate the tendency for creeping ideology with regard to museum exhibits and the deployment of pseudo-historical revisionism, facts, processes, and phenomena. These mechanisms can become a means of establishing science centres and museums as critical reconciliation, mediation, and recalibration contact zones for emancipatory interactions by diverse publics, visitors, and stakeholders. In addition, these dynamic contact zones could become critical catalysts for engendering new platforms, forums, and symposiums for initiating and sustaining mutual learning dialogue and respectful listening discourses between science communicators and practitioners from both the Global North and South.

These mutual learning interactions could engage on themes such as postcolonial perspectives on science communication history and innovation; multiculturalism and science and society; ethical dimensions in science communication; equitable and collaborative partnerships between science communication practitioners in the Global North and South; transformations in globally inclusive science communication practice, culture, and methodologies; and multidisciplinary frameworks for overcoming hegemonic thinking in science communication. Other potential emancipatory impacts include the possibility of science centres and museums contributing to the types of insightful activities that drive us to collective action and foundational issues of policy enrichment and the synergies of diverse recognitions. Science centres and museums could become empowering and engaging learning places that contextualise the fluidity of populations and demographic change at local and national levels. They should then reflect these inclusively in their practice, programmes, and methods, which change interactively as their communities become more diverse, through inculcating perspectives that drive collective innovation. The growth and proliferation of science centres/museum networks across both the Global North and South regions attest to their potential as a sector to deliver systemic transformation through inculcating decolonisation approaches as a means of 'centring' race and sociocultural inclusion good practice. These networks have a vital sector-wide leadership and peer–peer support (at both institutional and practitioner levels) and advocacy role to play in advancing diversity, equity, and inclusion in their membership institutions.

Conclusion: the brave new world for globally inclusive science communication – a level playing field for the Global North and South?

Science communication and public engagement discourses, paradigms, and platforms that enable the field to have a 'multi-lensed' understanding of the operations of the craft in diverse contexts and scenarios across the globe are important to building inclusive, engaged, open, and dynamic societies. These are highly critical mechanisms that will enable public participation and civic engagement in robust science policy decision-making and empowering progressive science and society agendas. Science communication that is articulated, demonstrated, and decolonised in ways and features that are understood within culturally relevant knowledge paradigms by globally inclusive publics and in multilingual settings creates an enabling environment for societal transformation. It furthermore engages the agency and autonomy of the publics, because through these processes, it can fully resonate with their life-worlds, leading to enhanced trust in and the vital reform of science communication institutions. The end goal is the creation of innovative forms of social cohesion in diverse societies and communities across the globe.

And why does this matter? It matters because in an age of nationalism and political populism, science communication has a much harder task to engender public trust in scientific facts, decolonised knowledge, and synthesised understandings of the nature of uncertainty and risk in scientific endeavour. This task will be much harder still if the field of science communication does not get its house in order and realise that the public do 'get it'. The public increasingly realise that science communication is not always the objective broker of science that it purports to be. This challenge of engendering public trust is just as much a pressing reality for science communicators in the Global South as it is for their counterparts in the Global North.

Science communication that continues to operate on the narrow premise of a one-dimensional Eurocentric framework, within a global scenario that is much more interconnected via a plethora of technology platforms, will become irrelevant and then redundant while the publics increasingly look elsewhere for their interface with science. The way forward lies in science communication adopting a transformative 'wide-angled' lens in order to better reflect the diverse global populations, contexts, and realities in both the Global North and the Global South. The many insightful lenses provided by the illuminating chapters in this book should contribute meaningfully to the framing and construction of this uniquely laudable goal.

This call to action for systemic change is as critical to the inclusive development of science communication in the Global South as it is in the Global North. It is imperative that practitioners in the Global South do not simply see themselves as passive victims of the exclusionary impacts of Eurocentrism in the field in both scenarios but, as has been amply illustrated by the inspiring contributions of some of their colleagues in this book, that they empower themselves, and with self-actualised agency, work to transform the field and its inclusive practices through leveraging the strengths and opportunities that exist in their local contexts and scenarios. A key strategy that will facilitate this empowerment is the solidarity that can be gained through Global South–South partnerships, networking, and shared action/advocacy, which would provide much needed peer–peer support, activism, and the sharing of good practices and resources. Strategic partnerships with peers in the Global North would also help facilitate these inclusive transformations and yield added-value innovative benefits for both sides, but as has been well documented in this book, they can only be of benefit if they are equitable and truly collaborative with a shared vision, ethos, and deliverables that are owned by and accrue to both sides – science communicators in the Global South as well as those in the Global North. The chapters in this book have also highlighted the exemplary transformation work being undertaken by some science communication institutions in the Global North and South regions over short-, medium-, and long-term timelines. It is vital that these inspiring developments are further amplified, scaled-up, shared, and disseminated across the field, in both regions, as an encouraging signpost of what sustainable change in diversity, equity, and inclusion can look like in the field across the globe.

One of the most sustainable ways forward in delivering transformative change is through the capacity building, training, and skills development of the many science communication actors in the field – academics, scholars, practitioners, researchers, and so on. Through good practice case studies, a number of chapters in this book illustrate how effective these measures can be both in the Global North and South, whether for science centre/museum staff and practitioners (Chapters 1 and 3), scholars and researchers (Chapter 4), or community-based practitioners (Chapter 12). It is to be hoped that the decolonised learnings and understandings, transformative knowledge, and inclusive practices from this book will be of use as a rich resource that provides critical, analytical, and globally relevant teaching, learning, and research resources to support the growing landscape of the many science communication/research/public engagement teaching, training, and capacity-building courses/modules across the globe. This is highly pertinent given the increasing popularity and growth of these courses in both the Global North and South regions, and their pivotal role in the capacity building and development of current and future generations of science communicators. The end-goal for these capacity-building institutions should thus be that of equipping science communicators with the requisite knowledge, tools, and social justice approaches to enable them to deliver science communication and public engagement that are directly informed by and fully reflect the contexts, realities, strengths, and challenges of their regions, localities, and diverse citizenry.

For their part, policy makers in the Global South have also been exhorted in this book to 'come to the table' and provide enabling, supportive, and resource-rich ecosystems in order for inclusive science communication to flourish in their regions of the world. The time for policy rhetoric and statements is over – they need to address the profound gaps that exist between their laudable policy frameworks and the reality of poor implementation of these transformative policies, where they exist, and to develop and implement them, where they do not.

Science communication policy frameworks in the Global North are undoubtedly more robust and better resourced than those in the Global South (Gascoigne and Schiele, 2020). However, given the need for far more enhanced delivery of inclusive science communication in the Global North, as articulated by the chapters in this book, this then begs the question: are these 'robust and better resourced policy frameworks' really fit for purpose for the advancement of diversity, equity, and inclusion in the field, in their localities, and contexts? And if not, why not?

Across the board, it is a given that science communicators in both the Global North and South are articulating a clear need for strategic leadership, drive, and sustainability for inclusive science communication in their regions, from policy makers, national governments, and enabling institutional frameworks. These leadership and strategic drives should also come from global multilateral organisations/institutions that operate in the scientific spheres as well as the many international STEM (science, technology, engineering, mathematics) scientific unions, associations, and academies that operate a membership framework that covers both the Global North and South regions. It is vital that these entities realise and act upon the pivotal leverage that they can bring to the advancement of globally inclusive science communication and the delivery of a level playing field for science communicators, researchers, practitioners, and other stakeholders across the inequities of the Global North–South divide.

The chapters in this unique volume have delivered a wide-ranging and comprehensive exposition of the state of the challenges of race and sociocultural inclusion in science communication across both the Global North and South regions of the world, and the inequitable ecosystems that enable exclusionary practices. They have in addition brought to the fore dynamic strategies, empowering perspectives, and transformative approaches for the delivery of progress, diversity, and sustainable systematic change. It is hoped that these enhanced understandings and innovative good practice signposts will guide the diverse stakeholders and practitioners in the field to amplify their efforts in delivering a radically inclusive and equitable ecosystem for science communication across the world – one that is fit for purpose for the 21st century and beyond.

References

Adichie, C.N. (2007) *Half of a Yellow Sun*, London: Harper Collins.

Alatas, S.F. (2006) *Alternative Discourses in Asian Social Science*, New Delhi: Sage.

Aman, R. (2018) *Decolonising Intercultural Education*, London: Routledge.

Asante, M.K. (2011) 'De-Westernizing communication: strategies for neutralizing cultural myths', in G. Wang (ed) *De-Westernizing Communication Research: Altering Questions and Changing Frameworks*, London: Routledge, pp 21–7.

Calvente, L.B.Y., Calafell, B., and Chávez, K.R. (2020) 'Here is something you can't understand: the suffocating whiteness of communication studies', *Communication and Critical/Cultural Studies*, 2: 202–9. doi: 10.1080/14791420.2020.1770823

Finlay, S.M., Raman, S., Rasekoala, E., Mignan, V., Dawson, E., Neeley, L. et al (2021) 'From the margins to the mainstream: deconstructing science communication as a white Western paradigm', *JCOM: Journal of Science Communication*, 20(1): CO2. doi: https://doi.org/10.22323/2.20010302

French, H.W. (2021) *Born in Blackness: Africa, Africans and the Making of the Modern World, 1471 to the Second World War*, np: W.W. Norton & Co.

Gascoigne, T. and Schiele, B. (2020) 'Introduction: a global trend, an emerging field, a multiplicity of understandings; science communication in 39 countries', in Gascoigne, T., Schiele, B., Leach, J., Riedlinger, M., Lewenstein, B.V., Massarani, L. et al (eds) *Communicating Science: A Global Perspective*, Canberra: ANU Press, pp 1–14.

Gunaratne, S.A. (2009) 'Globalization: a non-Western perspective; the bias of social science/communication oligopoly', *Communication, Culture & Critique*, 2(1): 60–82.

Hountondji, P. (ed) (1997) *Endogenous Knowledge: Research Trails*, Dakar: CODESRIA.

Howarth, F. (2018) 'Decolonizing the museum mind', AAM, Available from: https://www.aam-us.org/2018/10/08/decolonizing-the-museum-mind/

Maldonado-Torres, N. (2007) 'On the coloniality of being: contributions to the development of a concept', *Cultural Studies*, 21(2–3): 240–70.

Okere, T. (2011) 'Is there one science, Western science?', in R. Devisch and F. Nyamnjoh (eds) *The Postcolonial Turn*, Bamenda, Cameroon: Langaa and African Studies Centre, pp 297–314.

Olukoshi, A. and Nyamnjoh, F. (2011) 'The postcolonial turn: an introduction', in R. Devisch and F. Nyamnjoh (eds) *The Postcolonial Turn*, Bamenda, Cameroon: Langaa and African Studies Centre, pp 1–28.

Orthia, L.A. (2020) 'Strategies for including communication of non-Western and indigenous knowledges in science communication histories', *JCOM: Journal of Science Communication*, 19(2): A02. doi: https://doi.org/10.22323/2.19020202

R'boul, H. (2021) 'North/South imbalances in intercultural communication education', *Language and Intercultural Communication*, 21(2): 144–57. doi: 10.1080/14708477.2020.1866593

Rasekoala, E. and Orthia, L. (2020) 'Anti-racist science communication starts with recognising its globally diverse historical footprint', LSE Impact of Social Sciences Blog, 1 July, Available from: https://blogs.lse.ac.uk/impactofsocialsciences/2020/07/01/anti-racist-science-communication-starts-with-recognising-its-globally-diverse-historical-footprint/

Shoenberger, E. (2022) 'Museum Next – "What does it mean to decolonise a Museum?"', Available from: https://www.museumnext.com/article/what-does-it-mean-to-decolonize-a-museum/

Index

References to endnotes show both the page number and the note number (236n1).

A

ABCSSS islands, science communication in 188, 200–1
  and community-based engagement 197–200
  decolonisation of 190–1
  and dynamics of decision-making and co-creation 199–200
  'helicopter science' 191–2
  infrastructures of 189–90
  and local knowledge recognition 199
  long-term reciprocal collaborations building 199
  mental health care communication in 195–7
  and nature conservation research 193–4
  nature parks and fortress model in 192–3
  and public engagement 189, 191–4, 197–200
  and refusal to participate in research 200
  research and conservation agendas in 191–2
  research and funding 189–90
  transformations 197–200
Aboriginal and Torres Strait Islander knowledges 211
Adichie, C.N. 182, 241
Africa Scientifique (AS) Programme 6, 68, 69–70, 73
  and Afrocentricity 71–2
  learning by doing 74–6
  transformation for sustainable impact 72–4
Africa STI News 106
African Gong 7, 15, 66, 69, 77, 79, 81
African Science Literacy Network (ASLN) 106, 114
African Union 65, 101, 112, 177, 180, 184
Afrocentricity 3, 71–2, 119
agriculture 106, 108, 217
  and coloniality 88, 90
  in Nigeria 103–4
Airhihenbuwa, C.O. 42
ako Māori (culturally preferred pedagogy principle) 134
Alves, M.T. 161
Angelou, M. 150
anticolonialism 134, 136, 137, 140, 142, 151
Aotearoa New Zealand 130–2, 135, 138, 140–3, 141
  education sector 131
  Indigenous knowledge traditions 131
  knowledge governance in 132–3
  New Zealand Association of Scientists 133, 137, 139, 144
  New Zealand Science Review (NZSR) 131, 133–5, 137–9, 141–2
  see also Māori; mātauranga (Māori knowledge)
artefacts 8–9, 162–3, 170, 172, 241
artisans 214, 216, 218, 224, 226, 228–9, 232, 233
Aruba 189, 191, 193, 194, 195, 201
  see also ABCSSS islands, science communication in
Asante, M.K. 3, 71
Asia 180–1, 241
  minorities 35, 37, 39
  and SARS pandemic 37, 38, 42
  science centres and museums in 7
astronomy 134, 140, 213, 215, 226, 227
āta (growing respectful relationships) 135
audio recordings, as inclusion tool 57
Australia, and COVID-19 pandemic 39–42
  and Asian minorities 35, 39
  first cases 39
  Indigenous science 211
  issues with the state's response to 41–2
  local government areas (LGAs) 41, 42, 45
  second wave 39–40
  third wave 40–1
  33 Alfred Street tower 40
  and Victorian Department of Health and Human Services 40
Ayangunna, J.A. 122

B

Ba, H. 31
Bailey, B. 168
Balboa Park 152, 153
Baldwin, J. 163, 165
Balentina, X. 200, 202
Baloyi, C. 121
Banerjee, S. 232
Bang, M. 173
Barbican Centre 168
Battala publications 223, 233–4
rise of 223–30
role in knowledge democratisation 230–2
Bauer, M.W. 118
Baumgartner, R. 123
Bengal, colonial 222–3, 232–6
Battala publications *see* Battala publications
Bhadralok classes 224, 226–7, 230, 232, 236n1
Calcutta 224–5, 229, 232
and caste system 223–35
and hegemony and marginalisation 226–7
knowledge politics in 226–7
knowledge-sharing ecosystems 224–6
sociocultural contexts 224
Bengali language 226, 231
Biyela, S. 182
Black, Indigenous, and Peoples of Colour (BIPOC) communities 150, 152–4, 157, 158, 190, 195–7, 199
Black Africans researchers, and gender 66–8, 171
Black Lives Matter 184
Black movements, and art 168–9
Black Panther (film) 169, 170
blended language programming 56–7
Bonaire 189, 191, 194–7, 201
see also ABCSSS islands, science communication in
Bonaire National Marine Park 192
Bourdieu, P. 166
Brenchie's Lab 194, 199
Brua 195

C

Calcutta 224–5, 229, 232
Calcutta School Book Society 225
California Building 153
Cameroon 69, 70
Camit, M. 41
Canada
Royal Ontario Museum (ROM) 167
and SARS pandemic 35, 36–8
capacity building 12, 56, 243, 245–6
in the Caribbean 193
in New Zealand 131
in South Africa 66, 76, 77, 78
see also Africa Scientifique (AS) Programme
Carbonnier, G. 10
career advancement 70, 78, 109, 177, 180
Caribbean 94, 124–5, 200–2
see also ABCSSS islands
Caribbean, nature conservation and mental health in 198
dynamics of decision-making and co-creation 199–200
local knowledge recognition 199
long-term reciprocal collaborations building 199
and refusal to participate in research 200
Caribbean Netherlands Science Institute 198
CaribResearch 198
cartoon depictions, of SARS 37
caste, in colonial Bengal 228, 231
dominant groups 223
lower caste 223, 224, 226–7, 229, 231, 232–3, 234, 235
and privileges 224
upper caste 224, 225, 226–7, 230, 232–3
Castro-Gómez, S. 88, 89
chain of accessibility principle 57–8
Chickasaw Nation 151
Chilisa, B. 190, 193
China
ancient 11, 209
Haikou Village 125
Meng Hsi Pi T'an *see* *Meng Hsi Pi T'an* (Shen Kua)
and SARS pandemic 2, 36–9
Chivers, J. 115
Christoffel Park 192
Cisse, M. 120
citizen science initiatives 194, 199
citizens
engagement of 48, 110
and scientific knowledge 94, 102–3, 111
civil society organisations 13, 95, 168
civilisations 162, 169–70, 171–2, 173, 240
classes
in China 217
in colonial Bengal 223, 224, 226, 227, 230–4, 236n1
and STI ecosystem 67
Western-educated 100
climate change 108, 191
co-creative approach 32, 59, 116–17, 123, 124, 198, 199–200
colonialism
in ABCSSS islands 196
in *Black Panther* (film) 169, 170
and cultural imperialism 163–4
epistemic colonialism 87–9, 93
and exhibitions 161–3, 168, 169
impact on BIPOC 150
in *Indiana Jones* (film) 169–70
justification for 161–2, 171
in Mexico 85, 86, 96

and museums 149–50, 151, 152–3, 156–8, 164
in Nigeria 103
normativity 166–7
power structure 140
and science communication in Mexico 87–9
science education during 100
scientific 193
socio-psychological perspective 165
in South Africa 66, 67
Spanish architecture 153
see also Bengal, colonial
coloniality
and agriculture 88, 90
cultural 87–8
epistemic 87–9
of knowledge 7, 85–7, 89–91, 94
and scientific knowledge 85, 86, 89–92, 94, 172–3, 181, 209–10, 215, 234, 241–2
community-centred leave 157–8
Compton, L. 122
contextualisation 115–16, 118, 123, 243
copper-plate engravings 229, 233
Cornell, S. 132, 142
COVID-19 pandemic 2, 76, 115
in Australia 35–6, 39–42
and digital communication 121–2
and lockdowns 39–40, 42
in Nigeria 104, 110
and STEM education 20
vaccination development and delivery during 178
craftsmen 223, 226, 228–9, 230, 231–2
cultural archives 9, 161–2, 166, 168, 171
cultural heritage 156, 165–6, 211, 240
cultural imperialism 9, 162–4, 163–4, 166, 167–8
Curaçao 189, 191, 195, 197, 201
see also ABCSSS islands, science communication in

D

Darbel, A. 166
Das, S. 123
Dawson, E. 49, 222
decolonisation 87, 140, 147, 161–3, 167, 173, 239–43
and community-centred leave 157–8
guiding principles 154
of museums 10, 149–59, 241
of science communication 162, 181–3, 190–1, 201
deficit model 8, 36, 90, 110, 116–19, 125, 189, 201
democratisation
of knowledge, Battala publications' role in 230–2
of science 49–50
of scientific knowledge 209–10
Development Communications Network 106
Díaz, P. 90
Digdarshan 225
digital communication 120
discrimination 78n1, 196
and caste system 224
during pandemics 35, 37, 39, 42
and universal design principles 56
Du Bois, W.E.B. 162–3
Durham, J. 161
Durie, M. 133
Dutch Caribbean Nature Alliance 192
Dutta, U. 123

E

Ebola epidemic 105
e-health 197
Eken, H.N. 196
Empinotti, M. 109
epistemic injustice 92
epistemological orientations 173
ethics 9, 46, 97, 170, 186
Ethiopia 124
Etumnu, E.W. 124
Eurocentrism 23, 222, 239–42, 244
and colonial Bengal 225, 227
decolonisation and normalcy 167–8, 172, 210
and Global South 116, 118, 125
and Mexico 88, 95
and New Zealand 142
and Nigeria 101
and South Africa 67, 69, 71, 77
Exhibit B 168–9
exhibitions 57–9, 90, 154–5, 163, 167–9, 171, 242–3
Explainers 21, 219
co-designing demonstrations and programmes 29–31
and cross-departmental collaboration 29–30
definition of 22
experiences, impact of 26–8
habits of mind 29–30
learner-centred pedagogical approach and evaluation 31–2
relationships with science 22–6
and visitor-centred engagement 28
Ezcurra, E. 88

F

Fabian, J. 162
Fab-Ukozor, N. 124
Fafunwa, B. 105
Falade, B. 102–4, 108
Fanon, F. 165
Faraclas, N. 196
Ferdinand, M. 193

Finlay, S.M. 95, 118–19, 121
Fort William College 224–5
Foucault, M. 171
France 168–9, 210
French, H.W. 242
Fricker, M. 92–3
Friend of India, The 226

G

Gascoigne, T. 102, 106
gender 49, 50, 55, 58, 66–8, 73, 74, 101, 171–2, 177, 183, 233
Gerrard, J. 134
Ghana 69, 101, 176, 177
Ghosh, A. 230, 232
Global North 3–4, 9, 10, 13, 182
  and African challenges 177, 179, 183
  science centres in 48
  science communication epistemologies 118, 125
  training programmes in 68–9
Global North–South divide
  bridging through decolonisation 240–4
  complementary/alternative versus traditional medicines 182
  COVID-19 vaccines access 178
  funding divide 180, 181
  and globally inclusive science communication 244–7
  international collaboration partnerships 179–80
Global South 3–4, 9, 10, 13, 94, 115–16, 125–6, 177, 183
  deficit model in 119
  digital communication in 120
  and Eurocentrism 118–19
  inclusive science communication in 121–5
  inclusivity and public engagement 118–19
  participatory inclusive approach 117–18, 121
  public engagement approaches in 117–19, 120–1
  science and technology (S&T) policies in 119
  value of multimodality 124–5
  voices and communication actors, multiplicity of 122–4
globalisation 3, 7, 217, 239, 242
Goldman, A. 92
Gomez, M. 93
Goodhue, B. 153
Guan, D. 38
Guo, L. 209

H

Haikou Village 125
Halhed, N.B. 228
Hargreaves, T. 115
Harris, P. 134
health communication 104, 199
Hendy, S. 139
hermeneutical injustice 92, 93
Hidden Figures (film) 170–1
Hindu caste system *see* caste, in colonial Bengal
Hindu College 225
histories of science communication, significance of 210–12
  see also Meng Hsi Pi T'an (Shen Kua)
Hobson, J.M. 216–17, 220
Hountondji, P. 242
Hsu, E. 182
Hutchings, J. 134
Hyland-Wood, B. 42

I

identities 21, 32, 58, 60, 96, 154
  gender 49
  of museum 165
Igbo-Ukwu artefacts 241
Important Bird and Biodiversity Area 194, 199
In the Heart of Africa exhibition 167, 168, 169
inclusive science communication 239–47
  breaking the silos 50–1
  chain of accessibility 57–8
  and community partnerships 59–60
  and existing strengths 56–7
  in Global South 121–5
  inclusive learning environments 54–5
  and key people and stakeholders integration 58–9
  mission- driven intentionality 53–7
  and multilayered exclusion factors 119–21
  and public engagement 118–19
  strategies 51–3
  universal design principles 55–6
India 123, 222, 224–6, 234
  languages of 225, 228
  see also Bengal, colonial
Indigenous Australian Yorta Nation 95
Indigenous communication systems 120, 122, 124–5
Indigenous knowledge *see* Aotearoa New Zealand; Māori; mātauranga (Māori knowledge); Nigeria
Indigenous languages 8, 75–6, 120, 225
Indigenous peoples 93, 121, 124–5, 136, 139, 142, 150, 151, 153, 155, 182
innovation 53, 115, 122, 124, 179, 181, 183, 217
  in Africa 242, 243
  collective 243
  eco-innovation 180
  in New Zealand 130–1
  in Nigeria 100, 101–3, 106, 108, 111n5
  research-led 176–8

in South Africa 63, 66
technological 225
International Congress of Museums (ICOM) 164, 165
internationalisation 7, 180
intersectionalities 32, 60, 67–8, 142
Ishinaha-Shinere, S. 119
ITESO 93

J

Jackson, A.-M. 130, 134, 137–8
Jensen, P. 118
Jiang, T. 37

K

Kago, G. 120
Kalighat artists 230
Kapa-Kingi, Eru 135
Karmakar, P. 228, 233
kaupapa (collective philosophy principle) 134–5, 137, 141
Kenya 70, 76, 182
Khumalo, N.B. 121
kia piki ake i nga raruraru o te kāinga (socio-economic mediation principle) 134
Kim, Y.S. 217
knowers 91–2, 93, 136
knowledge
coloniality of 7, 85–7, 90–1, 94
governance of 131–2, 141–2
keepers 211, 213
production of 87–9, 100–1, 118, 142, 161, 172, 230, 242
systems 117–19, 121, 133, 140–1, 198, 211, 223, 226–7, 232
knowledge-making 223, 232–6
knowledge-sharing ecosystems 223, 224–6
knowledge-sharing practices 227, 235
Kolkata *see* Calcutta
Kontinen, T. 10
Kukutai, T. 134
Kumeyaay peoples 150, 153

L

Lagos Declaration and Call to Action on Science Communication and the Public Learning and Understanding of Science in Africa, The (2016) 65–6
land, Indigenous 141, 150, 153
language harmonisation 120
Latin America 7, 14, 56, 87, 94–5, 98, 124–5, 181
Lawal, O.A. 104
leadership 30, 32, 51, 102, 111n5, 122, 163, 177, 198–9, 246
learning
asset-based and community of practice approaches to 23
book-based 213, 214, 217
co-creative 124
in colonial Bengal 223, 225, 226
Design, Make, Play approach 21, 26
by doing 71, 74–6
experiences, inclusive 31, 32, 56, 57, 58–60
inclusive environment 24–5, 28, 52, 54–5
informal science learning 48–9, 53
mutual learning 239, 243, 245
science 21, 23, 24–5, 27
scientists and public, mutual learning between 117–18
as social practices 21
STEM 21, 30
learning environment, inclusive 24–5, 28, 52, 54–5
and community partnerships 59–60
key people and stakeholders integration 58–9
Leung, C. 37, 38
local communities 19, 54, 56, 59, 123, 192, 195–6, 201
local knowledge 88, 123, 197, 199
local languages 76, 102, 104, 109, 124–5, 197, 225
Lonetree, A. 149
Lozano, M. 94–5

M

Mac Donald, S. 192, 194
Maina, M. 105, 108
Māngai, R. 134
#Manosalacuenca 95
Manuel, R. 137
Māori 130, 141–2
Kaupapa Māori theory principles 134–5, 141
knowledge *see* mātauranga (Māori knowledge)
lunar and astronomical knowledge 140
Matariki 140
te reo Māori (the Māori language) 130, 131, 135, 136, 142
marginalisation 91, 116, 178, 226–7, 232
marginalised communities 73, 233–5
marine environment 194
markets 217, 231–3
Márquez, M.C. 123
Marsden, M. 143
Martín-Barbero, J. 92
mass media 37–8, 49, 122, 243
mātauranga (Māori knowledge) 130–1, 132, 142
epistemological scrutiny 138–40
governance 132–8
practice 134–6
and science governance practice 136–8
mathematics 69, 70, 73, 76, 210, 217, 226
see also STEM (science, technology, engineering, mathematics)

Maya Museum of Cancún 93
Maya peoples 156
Mayan cultures 93
Mayan Riviera 93
Medin, L. 173
Melbourne 39, 42
Memmi, A. 165
Meng Hsi Pi T'an (Shen Kua) 11, 209–10, 218–19
  jottings 213–16
  and Northern Sung society 216–18
  and science communication histories 212–13
  as a science communication text 215–16
mental health care communication, in ABCSSS islands 190–1, 195, 198, 200
  and BIPOC communities 199–200
  challenges to 195–6
  e-health and language in 197
Mental Health Caribbean (MHC) 195, 197
Mercier, O.R. 133
Mexico, science communication in 85, 86
  and epistemic coloniality 87–9
  and institutional framework 89–91
  recommendations for 94–6
  scientific knowledge and social epistemology 91–3
Mgbenka, R.N. 104
minority ethnic groups, and pandemics 35, 39, 42
Momen, H. 124
Monterey Bay Aquarium 54, 56
Moore, S. 60
Morgan, K. 137
Morris, M. 166–7
Mundy, P. 122
Museum of European Normality 161
Museum of Mayan Culture, Chetumal 93
Museum of Natural History, Mexico 90
Museum of Us 150, 158
  and Brandie 151–2
  and colonial legacy 152–4
  Community-Centred Leave 157–8
  and decolonisation 154–8
  Decolonising Initiatives in Action signs 155–6
  Eurocentrism and White supremacy disruption 157
  and land 150
  membership model 155
  and Micah 152
museums 4, 5, 7, 9, 19–20, 48, 53, 125, 242, 243, 245
  and authority 166
  case studies 167–9
  chain of accessibility 57–8
  co-designing demonstrations and programmes 30
  and cultural imperialism 164
  curators 166, 167, 169, 170, 172
  decolonisation of 150, 154–8, 162–3, 172, 241
  definition of 164
  ethnographic 163
  Eurocentric privilege 167–8
  existing foundation 56–7
  Explainers *see* Explainers
  and external experts 58–9
  gender inclusion in 171
  inclusion in 48–9, 50, 51, 52
  and inclusive community practice 29–32
  inclusive learning environment 54–5
  injustices in 93
  in Mexico 90–1
  mission of 20
  in New Zealand 130–1, 149
  and normalcy 164–5, 170–3
  NYSCI 21, 22
  and personal relationships 22–6
  and popular culture 169–70
  visitor-centred engagement 29
myths 102, 105, 138, 170, 188, 242

N

Nasarawa State University (NSUK) 107
National Advisory Council on Innovation 66
National Centre for Technology Management (NACETEM) 102, 111n2
National Institute of Water and Atmospheric Research (NIWA) Wellington Regional Science Fair, The 131
National Museum of Anthropology 91
National Office for Technology Acquisition and Promotion (NOTAP) 102, 111n2
National Science Foundation 19, 31
natural sciences 235
nature conservation, in Caribbean 190–1
  'helicopter science' 191–2
  nature parks and the fortress model in 192–3
  strategies and language in communication 193–4
Nepote, A.C. 86
Netherlands 167, 189, 191, 202, 205
New South Wales 42
New York Hall of Science (NYSCI) 20, 21, 32
  inclusive communication practices 29–32
  see also Explainers; museums; science centres
Next Einstein Forum (NEF) 108
Ngā Pae 137
Ngata, T. 140
Ngũgĩ, wa Thiongo 165
Nigeria 70, 100–1
  communication channels, diversification and popularisation of 109–10
  Federal Ministry of Science and Technology (FMST) 101, 102

funding adequacy 108
homegrown science culture 100, 103, 105, 106, 110, 111n1
hubris in science communication, addressing 110
Indigenous science communication in 107–10
local languages usage, in science communication 109
local scientists as role models 108–9
National Innovation System 102–3
policy makers and legislators 102
science, technology, and innovation (STI) agenda in 100, 101–3, 106, 111n2
science communication in 103–5
science communication initiatives in 105–7
teaching and practice of science, strengthening 107–8
Nigeria Centre for Disease Control (NCDC) 104
normalcy 160, 172–3
colonial normativity, enjoying 166–7
and cultural imperialism 163–4
and inaccurate narratives 170–2
as a multiplier of 'business as usual' inertia 164–5
popular culture and museums 169–70
North America 3, 7, 180–1, 200
Northern Sung society 216–18
Nyamnjoh, A. 240

O

Okeoma 241
Olukoshi, A. 240
online platforms 106, 122, 197
Oregon Museum of Science and Industry 60
Organisation of African Unity 101
Orientalism 3
Orientalism (Said) 161, 162
Orthia, L.A. 11, 95, 121, 188, 235
ownership
and mental health care 198, 199
of science 49, 102, 107
Oyewo, B.A. 122

P

Palmer S.E. 119
pan-Africanism 65, 69, 71, 100, 101
Panama-California Exposition 152–3
pandemics, and racial minorities 35
COVID-19 in Australia 35–6, 39–42
culturally sensitive communication 38
lockdowns, impact of 39–40
media coverage 37–8
participatory pathway 42–3
SARS in Canada 35, 36–8
see also COVID-19 pandemic; SARS pandemic
Papiamentu 197, 199
Pelican Conservation programme 194
Phalkey, J 222
PISKABON fisheries cooperative 194, 198
popular culture 161, 163–4, 169, 232
Porras, A.M. 123
printmaking practices 222–3, 226, 227–30, 232–4
private sector 57, 181, 217
Project Mātauranga 130
proverbs 74, 105, 124, 134
public communication of science 75, 85, 86, 92
public engagement 32, 95, 117, 239–40, 244
in ABCSSS islands 189–91, 193, 198–202
in colonial Bengal 233
in Global South 117–19, 120–1
in Nigeria 104
in South Africa 64, 69–70, 77–8, 81
public health 103, 104
public understanding 86, 103, 117, 118, 120
Pūhoro Academy 131
Pulido. G. 88

Q

Quijano, A. 87, 89

R

'Race: Are We So Different?' exhibit 154
Raj, K. 227
Ramos, A. 109
Rasekoala, E. 11, 95, 121
Rauika Māngai 134
refugees 12, 42, 58
religions 20, 49, 50, 196, 225
Reynoso-Haynes, E. 86
Robert Bosch Stiftung 107
Robert Bosch's Script project 111n5
Roediger, D.R. 163
role models 74, 108–9
Rosenthal, E. 38
Royal Ontario Museum (ROM) 167–8

S

Saba 4, 189, 191, 192, 195, 201
see also ABCSSS islands, science communication in
Saba National Marine Park 192
Said, E. 3, 161, 162, 164, 172
Samachar Darpan 225
San Diego Museum of Man *see* Museum of Us
Sanskrit 226
SARS (Severe Acute Respiratory Syndrome) pandemic 2, 36–9, 42
Schibeci, R.A. 119
Schiele, B. 102, 106
SciDev.Net 107
science, technology, and innovation (STI) policy 176, 181

in Nigeria 100, 101–3, 106, 111n2
in South Africa 63–6, 67, 68, 69, 71, 77
Science and Civilisation in China (Needham and Ling) 209
science centres 15, 20, 53, 54, 56–9, 165, 172–3, 242, 243, 245
decolonisation of 150, 154–8, 162–3, 165
existing foundation 56–7
gender inclusion in 171
inclusion in 48–9, 50, 51, 52
Science Communication Hub (SciCom Hub) 106
science education 239
in colonial period 100
informal 20, 52
and museums 19, 21
see also museums
in Nigeria 106, 107, 108
Science Museum Group 171
science museums *see* museums
Science Nigeria 106
scientific colonialism 193
scientific knowledge 7, 9–11, 49–50, 172, 225, 234
and citizens 94
and coloniality 85, 86, 89–92, 94, 172–3, 181, 209–10, 215, 234, 241–2
decolonisation of 241, 242
deficits in 117
democratisation of 209–10
institutionalisation of 89–90, 94
in *Meng Hsi Pi T'an* *see Meng Hsi Pi T'an* (Shen Kua)
and NYSCI 32
and social epistemology 91–3, 173
Western 227
Seleti, Y. 119, 121
Serampore Mission Press 225, 228
Sheda Science and Technology Complex (SHESTCO) 102, 107, 111n1
Shen Kua 11, 213–15
see also Meng Hsi Pi T'an (Shen Kua)
Simpson, L.R. 136, 137, 142
Sint Eustatius 189, 191, 195–6, 201
see also ABCSSS islands, science communication in
Sint Maarten 189, 191, 195, 201
see also ABCSSS islands, science communication in
Sint Maarten Nature Foundation 194
Sivin, N. 215
Siyanbola, W. 101, 102
Smith, L.T. 200
Sobane, K. 120, 122
social appropriation of knowledge 94
social media 75, 76, 104, 110, 170, 194, 243
socio-environmental problems 93, 95
Solomon, M. 91
South Africa 63–5, 70, 76–8
African Institute for Mathematical Sciences (AIMS) centres 68, 69, 70
apartheid 66–7, 78n1
colonialism in 66, 67
Department of Science and Innovation (DSI) 63–4, 66, 75
gender analysis 67
human resource capacity 65, 68, 76–7, 78
innovation in 63, 66
National Development Plan (NDP) 63
National Research Foundation (NRF) 63, 66–8
National Science Engagement Strategy (SES) Implementation Plan 64
National System of Innovation (NSI) 63
race and gender intersectionality analysis 67–8
racial analysis 67
science, technology, and innovation (STI) in 63–6, 67, 68, 69, 71, 77
science communication in 66–8
Science Engagement Strategy 66
socio-demographics and cultural characteristics 78n2
STISA 2024 65
'Year of Science and Technology' (YEAST) programme 64
see also Africa Scientifique (AS) programme
Spivak, G. 88
Statesman, The 226
STEAM (science, technology, engineering, arts, and mathematics) initiative 109
STEM (science, technology, engineering, mathematics) 19–20, 21–2, 30, 71, 78, 246
in ABCSSS islands 188, 189
and Eurocentrism 67
Explainers' experiences 27–8, 30, 32
and gender 52, 66, 68, 73, 170–1
professionals 211, 215
and race 68
in South Africa 66, 67
STEMM (science, technology, engineering, mathematics, and mātauranga) 131
Stewart, G.T. 137
strategic partnerships 71, 77, 181, 245
Sydney 39, 40, 41–2

T

Tallon, J. 138, 139
taonga tuku iho (cultural aspirations principle) 134, 135, 136
Te Koronga 138
Te Papa Tongarewa: Museum of New Zealand 130
Te Taiao exhibition 130
Te Tiriti o Waitangi (Treaty of Waitangi) 132, 135

Teaching and Research in Natural Sciences for Development (TReND) 106, 108, 111n5
techno-scientific knowledge 235
testimonial injustice 92–3
Théâtre Gérard Philipe 168
tino rangatiratanga (self-determination principle) 134, 135, 136, 142
Toronto 35, 36, 38
traditional medicine 107, 177–8, 182, 196
training programmes 68–71, 184, 199, 216, 227, 245
  for local health professionals 198
  locally led 193–4, 196
  STEM communication 72
  see also Africa Scientifique (AS) Programme
transformative inclusion 51

U

United Kingdom 168, 169, 171
United Nations Sustainable Development Goals (UN SDGs) 182
United States 5, 21, 22, 24, 48, 53, 54, 60, 157
  imperialism 151
  inclusion strategies in institutions 51
  and Kumeyaay 150
  and Mexico 89
  National Disability Authority's definition of universal design 55
  Public Understanding of Science (PUS) movement 86
  see also New York Hall of Science (NYSCI)
universal design principles 55–6
'unjust enrichment' biases 163

V

Vision Mātauranga (VM) 130–1, 133–4

W

Waitangi, Treaty of (ToW) 132, 135
Waitangi Tribunal 132
Wallerstein, I. 87
Walport, M. 49
Wang, Z. 123, 125
Washington Slagbaai National Park 192
Wekker, G. 161, 166, 167, 168
whānau (extended family structure principle) 134, 135, 138, 142
White Innocence: Paradoxes of Colonialism and Race (Wekker) 166
White supremacy 140, 151, 154, 157
Why, What, How framework (Cornell) 132–3, 142
Wiles, S. 139
Wilkins, C. 228, 233
women
  building capacity in South Africa 66, 67–8, 73, 78, 80
  and making knowledge visible 232, 234
  in Nigeria 102
  and science museums and centres 171–2
World Health Organization (WHO) 104

Y

'Year of Science and Technology' (YEAST) programme 64
Yopasa, M. 89
#Yoprefieroellago 95, 96

Z

Zambia 70
Zeng, J.-P. 209
Zhang, Y. 209, 214, 218
Zimbabwe 70
Zuo, Y. 214, 216
'Zwarte Piet' (Black Peter) 167